US Presidents and the Militarization of Space, 1946–1967

Number Nineteen: Centennial of Flight Series

Roger D. Launius, General Editor

A list of titles in this series appears at the end of this book.

US Presidents and the Militarization of Space, 1946–1967

Sean N. Kalic

Texas A&M University Press College Station

First edition

This paper meets the requirements of ANSI/NISO Z39.48-1992 (Permanence of Paper).
Binding materials have been chosen for durability.

Library of Congress Cataloging-in-Publication Data

Kalic, Sean N., 1970–
US presidents and the militarization of space, 1946–1967 / Sean N. Kalic. — 1st ed.
p. cm. — (Centennial of flight series ; no. 19)
Includes bibliographical references and index.
ISBN-13: 978-1-60344-691-4 (cloth : alk. paper)
ISBN-10: 1-60344-691-5 (cloth : alk. paper)
ISBN-13: 978-1-60344-697-6 (e-book)
ISBN-10: 1-60344-697-4 (e-book)
1. Astronautics, Military—United States. 2. Outer space—Strategic aspects. 3. Outer space—Government policy—United States. 4. Space security. 5. Presidents—United States—Decision making—History—20th century. 6. Cold War. 7. United States—Military policy. 8. United States—Politics and government—1945–1989. I. Title. II. Series: Centennial of flight series ; no. 19.
UG1523.K35 2012
358'.8097309045—dc23
2011034420

For Tracie and Katheryn, *the loves of my life*

Contents

Charts

Acronyms, Abbreviations, and Satellite Names

AAF	Army Air Force
ABM	antiballistic missile
ACDA	Arms Control and Disarmament Agency
AEC	Atomic Energy Commission
AFOS	Air Force Objective Series
ANNA	first geodetic satellite
APL	Applied Physics Laboratory
ARDC	Air Research and Development Command
ARPA	Advanced Research Projects Agency
ASAT	anti-satellite
BAMBI	Ballistic Missile Boost Intercept
BDA	bomb damage assessment
BMD	Ballistic Missile Defense
BMEWS	Ballistic Missile Early Warning System
BOMI	bomber missile
CIA	Central Intelligence Agency
CIC	Critical Issues Council
COMSAT	communications satellite
DCI	Director of Central Intelligence
DDEPL	Dwight D. Eisenhower Presidential Library
DEW	Distant Early Warning
DIA	Defense Intelligence Agency
DoD	Department of Defense
EMP	electromagnetic pulse
FOBS	Fractional Orbital Bombardment System
FY	fiscal year
GPS	Global Positioning System
GMC	Guided Missile Committee
GWUNSA	George Washington University, National Security Archive
HSTPL	Harry S. Truman Presidential Library
ICBM	intercontinental ballistic missile
IGY	International Geophysical Year
IOC	initial operational capability
IR	infrared
JFKPL	John F. Kennedy Presidential Library
LBJPL	Lyndon Baines Johnson Presidential Library
MIDAS	Missile Defense Alarm System
MOL	Manned Orbital Laboratory

NACA	National Advisory Council on Aeronautics
NASA	National Aeronautics and Space Administration
NASAHO	National Aeronautics and Space Administration, History Office
NASC	National Aeronautics and Space Council
Nimbus	weather satellite
NRL	Naval Research Laboratory
NSAM	National Security Action Memorandum
NSC	National Security Council
NYT	*New York Times*
OSANSA	Office of the Special Assistant for National Security Affairs
OSD	Office of the Secretary of Defense
OST	Outer Space Treaty
PCIAA	President's Committee on Information Activity Abroad
PLBJ	Papers of Lyndon Baines Johnson
PPDDE	Personal Papers of Dwight D. Eisenhower
PPJFK	Personal Papers of President Kennedy
R&D	Research and Development
RD&E	Research, Development, and Engineering
Relay	active communications satellite
ROBO	rocket bomber
SAB	Science Advisory Board
SAC	Strategic Air Command
SAINT	Satellite Interceptor
SAMOS	Satellite and Missile Observation System
SCORE	Signal Communication by Orbital Relay Equipment
SIGNIT	signals intelligence
SLBM	submarine-launched ballistic missile
SPAD	Space Patrol Active Defense
SSBN	ballistic missile submarine
TCP	Technologies Capability Panel
TIROS	Television Infrared Observation Satellite
Transit	navigation satellite constellation
UN	United Nations
USIA	United States Information Agency
USAF	United States Air Force
USN	United States Navy
VANGUARD	US Satellite for IGY run by USN
Vela Hotel	nuclear detection satellite
WDD	Western Development Division
WHO	White House Office

Acknowledgments

The completion of this book project brings with it the need to thank a diverse group of people who have assisted me along the way. First and foremost, I need to offer a hearty thank you to Donald J. Mrozek, who served as my mentor. In addition to providing advice, words of encouragement, and guidance, he always made sure that I took time to appreciate the finer points of life. For his expert guidance and hospitable friendship, I will always be grateful.

While at Kansas State University, I had the pleasure of working with an outstanding group of professors who offered insightful advice and helpful comments. Thank you to Dr. Jack M. Holl, Dr. Mark Parillo, Dr. David Stone, and Dr. Joseph K. Unekis for seeing the potential of this book.

In addition to my professors, I must thank several institutions for providing financial support for my research: the Harry S. Truman Library Foundation, the John F. Kennedy Library Foundation, the Lyndon Baines Johnson Library Foundation, the Institute for Military History and Twentieth-Century Studies at Kansas State University, and NASA's History Office. While researching at various archives and presidential libraries, I had the privilege of working with many dedicated and generous individuals, including Stephen J. Graber, Colin Fries, and Jane Odom from NASA's History Office, Dennis Bilger from the Harry S. Truman Presidential Library, David Haight and Barbara Constable of the Dwight D. Eisenhower Presidential Library, Sharon Kelly from the John F. Kennedy Presidential Library, and Tina Houston from the Lyndon Baines Johnson Presidential Library.

In addition to these fine institutions and their dedicated people, I also need to thank Mary Lenn Dixon, editor-in-chief, Texas A&M University Press, for her support and encouragement throughout the process. Copy editor Chris Dodge, I truly appreciate your insightful assistance. Lastly, I need to thank all my colleagues in the Department of Military History at the US Army's Command and General Staff College, especially Dr. James H. Willbanks, for your camaraderie and support during this project. It is a privilege to work with professionals like you.

I have also benefited from the support and love of friends and family. To the members of Blue Valley Yacht Club, especially Henry Otto, thank you for providing me with countless opportunities to hone my sailing skills and make sure that I did not suffer the typically pasty skin of an academic. To the crew of *Line Drive,* Randy Zelenka and Ken Hays, thank you for keeping me humble by always allowing me to serve as your bowman. It has been a great few years on the "fastest Antrim 27 in the Midwest." In addition to this colorful cast of sailors, I must also acknowledge the love and support of my family. To my parents, Denny and Sandy Kalic, thank you for always believing in my abilities and pushing me to achieve my goals. Your love and support can never be repaid, but it has always been deeply appreciated. I am also indebted to my brother Jon Kalic and his wife Jolee and their children, Cameron and Jillian, for providing humor and support during years of work on this book. Thanks for helping me keep everything in perspective. I am also fortunate to have the support of my in-laws, John and Peggy Brine, who not only allowed me to move their daughter to faraway Kansas, but also always provided words of encouragement.

Finally, I could not have completed this project without the love of my wonderful wife, Tracie. She served as my editor, travel companion, financier, best friend, soul mate, and chief supporter. Thank you for putting up with me and providing me with a wonderful life. I have also had the joy and pleasure of our daughter Katheryn, who came along well into the project, and has been a great distraction. I love you both and I could not have done this without my ladies. And I need to thank our dogs Nigel and Murphy who always seemed to know when I needed to take a break and go for a walk.

US Presidents and the Militarization of Space, 1946–1967

Introduction

The spectacular shooting down of a nonresponsive spy satellite by the US Navy's USS *Lake Erie* on February 19, 2008, and the Chinese interception of a Fengyun-1C weather satellite on January 11, 2007, focused the attention of the American public, space experts, and the US military on the strategic importance of outer space. These events renewed discussions in the international community as well as the US government on the militarization and the potential future need to weaponize space. During 2005, a reporter for the Fort Worth *Star-Telegram* had quoted the then commander of the Air Force Space Command, Gen. Lance Lord, about the use of weapons in space. Lord affirmed, "[W]e believe in the peaceful use of space."[1] Although this quotation from General Lord may seem contradictory to the actions taken by the United States and China in 2008 and 2007, his phrase outlines the fundamental idea that led US presidents to define a national space program between 1946 and 1967. More specifically, Lord's words provide a catchphrase for an interpretation of the simultaneous push for the militarization of space and the banning of weapons from space by US presidents between 1946 and 1967.[2] Basically US presidents from Truman to Johnson did not race the Soviet Union to arm the heavens with weapons, but believed in the peaceful use of space for military and civilian purposes and worked to construct a national space policy that supported the civilian and non-aggressive military uses of space.

The debate generated in by the events of 2007 and 2008 was not new, but rather the continuation of a national policy debate that the United States has revisited every decade since the 1950s. As nations become increasingly reliant on satellites for radio, television, and services that range from weather and navigation information to advanced military applications, space becomes a vital national security issue that demands attention, discussion, and forethought. Luckily, policy makers, analysts, and military officers have a rich archive of space history to inform them. As early as 1945 the United States began to research and construct a basic framework that would eventually yield a national space policy that contained both military and civilian space programs. The framework developed in the initial years of the early Cold War continues to serve today.

The rich and dynamic history of the process by which the United States came to define a national space policy, one built on the premise that space needed to be maintained as an open scientific frontier for the benefit of all nations, is often overshadowed by certain of its facets: intriguing spy satellite programs, historic manned space flights, and presidential initiatives. The process by which successive US presidents from Eisenhower to Johnson presided over the development of a national space policy needs attention, as the foundations established in the period 1954–1967 serve as the parameters for the current space policy debate.

Before this debate can progress, it is important to understand the past and recognize the roles of Presidents Eisenhower, Kennedy, and Johnson in forging the "peaceful use of space" policy eventually codified with the signing of the Outer Space Treaty in 1967. While continually and sincerely putting forth a simple message in favor of "peaceful usage," Eisenhower, Kennedy, and Johnson also recognized the need for the military use of space, however.

These two ideas, apparently mutually exclusive, served as guideposts by which these three presidents strove to define a national space policy that provided for the militarization of space, while also affording the civilian and scientific exploration of space for the "benefit of all mankind." This was by no means a simple matter in the context of the Cold War. Understanding the process by which the United States defined and built a national space policy between 1945 and 1967 can provide a new generation of Americans with the background needed to define a space policy for the twenty-first century.

As the United States considers a national space policy for the twenty-first century, it is imperative to understand the intellectual ideas and paradigms that shaped history. Although historians have written award-winning histories of the space age, there exists a gap in exploring the continuity between US presidents as they worked to construct a viable space policy. In an effort to understand this policy-making process, it is important to stay focused on the continuity of the relevant issues and not be sidetracked by the history of specific satellite programs or other national security policies that may intersect or run parallel to the story. Because of the focus of this study specific programs such as Corona, Vanguard, and Dyna-Soar—and policies such as Open Skies and New Look—will not be discussed in great detail, but only with a view to how the respective presidents viewed them in relation to the development of a national space policy. This study is not a comprehensive history of the space age, but rather the history of the ideas and actions that shaped Eisenhower's, Kennedy's, and Johnson's views of the uses of space.

Background

The birth of the space age has long been identified by space policy historians with the October 4, 1957 launch of *Sputnik* by the Soviet Union, and with the subsequent efforts by the United States to place man-made objects into orbit.[3] Ignoring the critical first decade of US government interest in the military and civilian uses of space overlooks the fundamental intellectual work done under Harry S. Truman's administration by key space advocates from the US Navy (USN), US Air Force (USAF), and the RAND Corporation.

For both the United States and the Soviet Union, space became a critical component of the Cold War arms race from its onset. The history of the civilian space program has overshadowed the history of the quest by the US government to define a space policy founded on the ideal of non-aggressive military uses of space.[4] Simply, the early history of US government efforts to militarize space has received less attention by writers. There are fine historical studies detailing the military's interest in the use of space, by authors such as Paul Stares, Norman Friedman, Donald Baucom, Jeffrey Richelson, and David Spires, but there is no single study covering the evolution of the efforts by US presidents to build a policy focused on the use of space for peaceful purposes.[5] Building on the foundation established in the period 1945–1952, Presidents Eisenhower, Kennedy, and Johnson crafted a space policy that allowed for non-aggressive military missions, while also working with the international community to ensure that space remained an open and weapons-free frontier for all nations. The decisions by these three presidents to use space for military as well as civilian purposes were not made in a vacuum. Rather all three embraced the use of space for reasons of national security, international prestige, and scientific and technological research, all areas of heightened concerns during the Cold War.

The very concept of non-aggressive military use of space is important to consider. From the first space-related feasibility study it produced, the RAND Corporation consistently defined satellites as having significant non-aggressive military uses, meaning that satellites could perform functions such as photoreconnaissance, weather data collection, and communication missions without threatening the world below. Although the US Air Force identified the need for space weapons as the space race developed between the United States and the Soviet Union, RAND and later Eisenhower, Kennedy, and Johnson strongly advocated the non-aggressive militarization of space, while vehemently rejecting the development and deployment of space-based weapons. These three presidents believed that space provided an opportunity to improve the stability of the security environment, while also demonstrating to the international community the peaceful intentions of the United States. For the next four decades, both the Soviet Union and the United States covertly and openly competed to maximize the advantages of space.

The Presidents and Space

The history of US efforts to militarize space and at the same time work to ban weapons from it reveals both significant continuity and much variation in the actions and policies of the individual presidents.[6] Truman remained personally detached from the work on early ideas about the use of satellites and space systems, while members of the RAND Corporation and USAF officers defined the basic uses of military satellites. Future presidents built onto this foundation for the use of non-aggressive military satellites. The Truman era also saw the birth of the idea that eventually led his successors to push for an international ban on the stationing of weapons in space. Eisenhower built on the foundation established under Truman's administration and personally supported the development of both a civilian and military space program.

Unlike Truman, Eisenhower recognized the significant national security contributions that non-aggressive military satellites could provide the United States. He also saw the importance of advancing the US civilian space program as a demonstration of US commitment to the peaceful use of space. These ideas carried over into the administrations of Kennedy and Johnson.

President Kennedy emphasized the role of NASA by embracing the capabilities of the civilian space program to further the US agenda on the peaceful use of space. Despite strong criticism for his decision to emphasize NASA's programs over those of the military, Kennedy further advanced the use of non-aggressive military satellites as developed and authorized under Eisenhower. In breaking with Eisenhower's support of two officially separate space programs, Kennedy supported the concept of integrating the military and civilian space programs of the United States into one national program.

After Kennedy's assassination, President Johnson continued to support Kennedy's idea of a single national space program. Like Kennedy, though, Johnson recognized the importance of both the military and civilian use of space. In 1967 Johnson oversaw the ratification of the Outer Space Treaty (OST) that secured the US objective of banning weapons from space. However, even as he led the international effort to ban weapons from space, Johnson supported the development of ground-based anti-satellite systems and ballistic missile defenses for the national security of the United States. While securing the role of world leader in the peaceful use of space, Johnson also enhanced the militarization of space by the United States.

Ultimately, Eisenhower, Kennedy, and Johnson worked to ensure that space did not become another weaponized frontier of the Cold War. Although the history of the space race between the United States and the Soviet Union has largely been explained by historians as a product of national security concerns, national prestige considerations, or a product of international rivalries, these interpretations have overlooked the continuity of the refrain from US presidents that the United States supported the peaceful use of space. Remaining consistent with this belief, Eisenhower, Kennedy, and Johnson supported the capabilities and programs of the both the military and civilian space programs of the United States and their policies entailed both militarizing space (see definition below) and banning weapons from space. The fundamental work completed during the Truman presidency contributed greatly to these efforts.

Definitions

In order to understand the evolution of US space policy during 1946–1967, working definitions of the terms "militarization of space" and "weaponization of space" are critical. Militarization of space came to be regarded as the use of space-based systems to collect, gather, and disseminate photographic intelligence, communications data, weather data, signals intelligence, and strategic reconnaissance. Note that the term "militarization" here is not synonymous with the use of military force or the aggressive use of space weapons. Rather, in the evolution of US space policy, Presidents Eisenhower, Kennedy, and Johnson came to use the term "militarization of space" to convey their interest in the use of space for non-aggressive military purposes. Throughout the evolution of US space policy, presidents consistently believed that the US military space program supported their objective to use space for peaceful purposes.

In contrast to their definition of the militarization of space, Eisenhower, Kennedy, and Johnson defined "weaponization of space" as the use of space-based systems to defend against the use of other space-based weapons or to deny an enemy access to space, the use of space-based weapons to target terrestrial sites, and the use of space weapons to destroy an enemy's space-based assets. For the period 1957–1967, the term "weaponization" was synonymous with the aggressive use of space and space-based systems. Eisenhower, Kennedy, and Johnson associated space-based anti-satellite (ASAT) systems, ballistic missile defenses (BMD), and orbital bombardment systems as the standard weaponization concepts that threatened space as an open frontier.

These presidents believed that their support of research and development for ASAT and BDM systems did not violate their pledge to keep to the peaceful use of space because the systems were ground-based and not space-based. They argued against the use of space-based weapons while they supported the development of ground-based defensive systems capable of intercepting space-based weapon systems and thereby preserving the openness of space for all nations to explore.

Chapter 1: Establishing the Foundation for the Militarization of Space, 1945–1952

The years 1945–1952 are not usually considered part of the space age, yet these years witnessed the intellectual beginning of the US military space program and the origin of US policy to promote space as a peaceful place for nations to explore. In fall 1945 through spring 1946, the US Navy (USN), the US Army Air Force (AAF), and the Douglas Aircraft Company's Project RAND began initial technical feasibility studies of the military usefulness of satellites. These inquiries established the theoretical foundations for the later development of the US military and civilian space programs.

At the close of World War II, the USN and USAF competed to win control of a mission not yet described or defined.[1] Based on the technology of Germany's V-2 rocket, both services examined the idea of rockets and satellites for military purposes.[2] Despite interest in satellites among the military services, President Truman himself devoted no attention to space systems. Instead, he favored aeronautical research and development programs. Without presidential interest, engineers and social scientists of the RAND Corporation, under contract from the air force, worked steadily to define the military use of satellites and the civilian benefits of space exploration between 1945 and 1952. By 1945, the space age had begun.

Considering the Feasibility of Satellites

Before the publication of RAND's groundbreaking study on satellites for the air force, the navy led the US military's initial interest into the use of satellites.[3] On October 3, 1945, members of the Naval Research Laboratory (NRL) and Bureau of Aeronautics conducted preliminary studies on the technical feasibility of satellites and established the Committee for Evaluating the Feasibility of Space Rocketry.[4] Through an initial contract with North American Aviation and the Guggenheim Aeronautical Laboratory, the navy funded the first full-fledged technical feasibility study for a human-made satellite.[5]

While it was promising for military application, the navy recognized that the program's estimated cost of five to eight million dollars would be beyond its financial resources and budget allocations.[6] In an attempt to distribute the costs of a satellite, Cdr. Harvey Hall of the navy's Bureau of Aeronautics proposed a joint USN-AAF satellite research program, raising the idea in a meeting with Army Air Force generals H. J. Knerr, H. W. McLellan, and W. L. Richardson in March 1946.[7] The generals agreed to discuss the project with the AAF director of research and development, Maj. Gen. Curtis E. LeMay. Stunned, LeMay rejected the navy's proposal for joint research on satellites outright and extended a contract to the Douglas Aircraft Company for an AAF feasibility study on an earth-orbiting satellite. Seeing an opportunity to advance AAF efforts to gain independence, LeMay believed that the AAF, not the navy, should control all aspects of air operations, even if this came to include satellites and space systems. Subsequently LeMay established an AAF satellite feasibility program within the Douglas Aircraft Company's Project RAND, which later evolved into the RAND Corporation. In essence, the air force program rivaled the navy's study.

In the years immediately after the Second World War, the United States already identified itself as the leader in aeronautical research and development, and Project RAND evolved from AAF general H. H. Arnold's belief that America needed to maintain and strengthen its scientific and technological edge in the Cold War. Arnold feared that US research and development would wither after World War II. Reacting to the navy's proposal and striving to maintain the relationship between air force and civilian scientists, LeMay charged the engineers of Project RAND with assessing the feasibility of a satellite.[8] With the steadfast belief of air force leaders that the future of military operations rested upon the strength of America's strategic airpower, leaders such as LeMay foresaw satellites as potentially useful for the air force.[9]

On May 2, 1946, engineers in Project RAND presented their report entitled "Preliminary Design of an Experimental World Circling Spaceship."[10] Douglas Aircraft Company engineer and report author Louis Ridenour outlined the significance of the satellite vehicle and the potential military applications of satellites for the AAF. According to Ridenour, satellites could feasibly correct the flight paths of missiles or rockets along their ballistic trajectories, provide bomb damage assessment, and supply weather observation over targets. As for a satellite's offensive capability, Ridenour simply speculated that satellites could be used as missiles that could be de-orbited over a specified target.[11] Despite the potential embodied in Ridenour's observations,

Gen. H. H. Arnold with Gen. Curtis E. LeMay (back right) visit NACA's Aircraft Engine Research Laboratory, November 9, 1944. NASA, courtesy of nasaimages.org

the technical, political, and economic climate of the late 1940s precluded the AAF from investing large sums of money into satellite research and development. The pressing issues of rearmament and containing the Soviet Union garnered Truman's attention and that of his national security staff, and they effectively ignored the use of satellites and space research. While USAF kept funding studies in an effort to position itself as the nation's future space force, their research establish the foundation for the militarization of space.[12]

Considering the future of satellites, RAND's engineers initially stated that, although the exact design was not yet known, two things seemed clear:

- A satellite vehicle with appropriate instrumentation can be expected to be one of the most potent scientific tools in the twentieth century.
- The achievement of a satellite craft by the United States would inflame the imagination of mankind, and would probably produce repercussions in the world comparable to the explosion of the atomic bomb.[13]

Ultimately the group of engineers surmised that a man-made satellite would be of great value in presenting the United States as the world's technological leader, a vital asset in the emerging Cold War with the Soviet Union.[14]

To highlight the military value of satellites, Louis Ridenour identified reconnaissance, navigation, intelligence gathering, communication, and targeting as the five basic military functions a satellite could eventually perform. Over the next twenty-years, these five functions became the basic missions for US military satellites. The engineers involved in RAND's initial feasibility study, also known as SM-11837, concluded that the United States possessed the technology to build a satellite.

Notwithstanding the promising research done by RAND, the Truman administration stayed focused on airpower and atomic weapons as they believed these systems were vital elements in the US military's force structure in the emerging struggle against the Soviet Union. As such, the major USAF research and development programs focused on airpower and atomic weapons, while satellites received little recognition.

Concern over communist expansion in the early Cold War drove Truman to focus on the development of national security strategies to contain communism rather than on a satellite program that might not have fruitful military applications. In the late 1940s, atomic weapons, conventional military hardware, international alliances, and rebuilding Germany and Japan received the bulk of Truman's attention as he struggled to build a postwar national security strategy for the United States. Although the navy and air force expressed interest in the technical feasibility of satellites, and had even outlined potential military uses, both services chose to invest in nuclear and conventional forces that they believed were of more immediate relevance to the national security of the United States.[15]

In the first decade of the Cold War, the air force and the navy needed to justify defense expenditures in increasingly tight budgets.[16] The navy lost its fight for a proposed strategic super-carrier, the USS *United States,* while the air force received authorization for its intercontinental B-36. The struggle between the services highlighted that neither service could afford to propose a seemingly farfetched R&D plan such as building satellites or exploring outer space when they needed to focus on rebuilding the conventional and nuclear forces of the United States.[17]

As the international security climate of the late 1940s intensified, it seemed even more unlikely that the Truman administration would support the development of satellites amid concerns over the status of US airpower vis-à-vis the Soviet Union, which prompted Truman to appoint a committee to investigate the strategic military balance in 1947. Truman's Presidential Air Policy Commission, headed by the US government's advisor to the US mission to the United Nations, Thomas K. Finletter, focused on the need of the United States to improve its airpower and aeronautical technologies based upon intelligence reports indicating that the Soviet Union was significantly increasing it airpower capabilities.

In open contradiction to the intelligence reports, navy Rear Adm. D. V. Gallery stressed in a letter in October 1947 to Air Policy Commission member John A. McCone that he believed "Russian capabilities are below those of the United States."[18] One month later, however, E. E. Partridge (USAF major general and acting deputy chief of staff for operations), B. L. Boatner (USAF brigadier general and air force liaison officer to the Air Policy Commission), and S. Paul Johnson (executive director of the Air Policy Commission) raised concerns about the technological balance between the United States and Soviet Union in the development of air power and missiles. These men argued that "Russia's developments in air power and missiles are probably further advanced than the United States."[19] The polarized opinions demonstrated the lack of hard intelligence possessed by the United States in the immediate years after the Second World War.

Reinforcing the view that there was inadequate intelligence on the Soviet Union, E. C. Sweeney, legal advisor to the Air Policy Commission, told navy officials in a meeting with the navy's Bureau of Aeronautics in September 1947, "[W]e have little specific knowledge of Russian developments."[20] The United States needed to keep informed of the expanding military capabilities of the Soviet Union. For the next several years, the United States worked at using various endo-atmospheric methods to collect vital strategic intelligence.[21] The Soviet Union's surprise detonation of its first atomic bomb in 1949 demonstrates the lack of intelligence possessed by the United States on the military capabilities of its Cold War rival.

Beyond the need for more intelligence, the Air Policy Commission gauged the Soviet Union's capability to develop, construct, and deploy missiles and atomic weapons as compared to the United States. In their preliminary report the committee members assumed the Soviet Union would have an atomic weapons capability by 1953, reasoning that if the Soviet Union had begun a program in 1943, it would likely achieve desired results within ten years. In order to counteract any concern over the Soviet Union's scientific capabilities, the commission's members noted that a ten-year program was "three times as long as it took the United States to complete the same job." The commission believed that their ten-year estimate was realistic.[22]

Despite the estimate, the former chairman of the War Production Board, Donald M. Nelson, testified before an executive session of the Air Policy Commission on October 23, 1947, stating that "the Russians will have the bomb in two to three years."[23] As with the discrepancy over the time needed by the USSR to develop guided missiles, the uncertainty of intelligence produced significant differences in the advice given to Truman by military and civilian experts on the exact date when the Soviet Union would detonate its first atomic bomb. While serving as a stark reminder of how little intelligence the United States actually possessed on the Soviet Union's atomic bomb project, this reinforced the need for the United States to gather intelligence on the Soviet Union's other rapidly expanding military capabilities.

The loss of America's atomic monopoly made many Americans, including Truman, LeMay, and the head of the State Department's Policy Planning Staff, Paul H. Nitze, concerned about rearming. Nitze and the members of the State Department's Policy Planning Staff recommended that the United States rebuild its conventional forces.[24]

Although the Air Policy Commission recommended that the United States invest in advanced research and development programs, the commission's members completely ignored the prospective role of satellite and space research in US national security. The members of the commission advised Truman that the United States build more civilian and military aircraft as well as invest in advanced aeronautical research and development programs.[25]

Even though the air force remained interested in satellites, it assigned the highest priority to rebuilding its strategic air power. Concerning the nation's status as the world's premier airpower, Maj. Gen. Gillispie Craigie, air force deputy chief of staff for materiel, stated in October 1947: "Our production is a mere trickle and our once vast fleet of airplanes no longer exists, and those aircraft remaining are rapidly becoming obsolete."[26] Alarmed at the decline of US aircraft production capability, despite the need for new bombers, fighters, and research aircraft, Craigie advocated federal investment in a "long range research and development program" designed to strengthen the American aircraft industry.

The National Advisory Committee for Aeronautics (NACA), which had been established in 1915 by Congress as a government organization to pursue aeronautical research, agreed with Craigie, and in November 1947 asserted that the United States needed more "trained technical personnel, advanced high-speed wind tunnels, and proper facilities for the development of aircraft and missiles" to remain the international leader in aviation.[27] Despite cursory talk by some military personnel about the need to use science and technology to develop "radically different weapons," the committee never mentioned satellites once in its discussions.[28]

Even after the promising feasibility studies done by the RAND Corporation, the air force identified serious technical hurdles to developing satellites for military purposes. Primarily, the lack of an adequate rocket booster remained a paramount concern for the air force. The captured German V-2, the subsequent American versions, and smaller research rockets of the army and navy (Viking and Aerobe) lacked the capability to lift the satellite payload envisioned by RAND's engineers. As already noted, the air force had staunchly committed itself to building a modern strategic bomber force. These two variables significantly eroded the initial enthusiasm of the air force to invest in a costly military satellite program.

While the air force and RAND worked on the feasibility of satellites, the army and navy sought to solve the technical issues associated with the development and production of missiles that could also function as boosters for satellites. Despite LeMay's rejection of cooperation with the navy on the feasibility of satellites, as early as 1945 the army and navy had pieced together a Guided Missile Committee (GMC), led by industrial chemist Bradley Dewey, to work on the development of both short-term and long-term missile programs.[29]

The GMC submitted its short-term and long-term reports to the Joint Chiefs of Staff on August 10 and November 21, 1945, respectively. Although the members of the committee faced infighting and interservice rivalry over missile designs and specifications, the committee succeeded in establishing performance parameters upon which several missile programs evolved in the late 1940s. The difficulty of reaching a consensus among the GMC members prompted America's guided missile expert Lawrence R. Hafstad to caution about the quick development of guided missiles. As director of research at the Johns Hopkins Applied Physics Laboratory (APL), Hafstad warned that the path to the development of guided missiles involved a complex web of "aerodynamics, electronics, thermodynamics, metallurgy, and chemistry." Furthermore, Hafstad offered that the United States needed to master the production of "radars, electronics, servo-mechanisms, gyros, computers, and propulsion" before an operational missile could be fielded.[30] From the onset, civilian as well as military scientists warned against expecting the research and development phase for rockets and missiles to be short.[31]

A major change in the status of the air force deeply affected the argument over the development of space forces. While a segment of the scientific and military community worked toward building missiles in 1945 and 1946, the officers of the Army Air Force pushed for independent status. The National Security Act of 1947 gave USAF independence. The emergence of an independent air force brought with it the need to redefine the roles and missions of US military services, including the prospective realm of missions in space.

This readjustment of the roles and missions of the armed services contributed to the lag in America's missile and rocket development. Historians have tended to overlook the vast amount of time, energy, and resources marshaled by the air force to define itself as an independent service. While engaged in a bitter and public debate with the navy over strategic missions, the air force favored the procurement of the B-36 and follow-on jet bombers. With the vast majority of the air force focusing on the development of its postwar air power, a cadre of engineers and social scientists from the RAND Corporation focused instead on refining the military roles and missions of satellites.

In the 1950s, the air force established a foundation for its future claim to be the nation's space force by building on the "agreed upon primary function" of the air force outlined in USAF-USN meetings in Key West in 1947.[32] According the agreement worked out in Key West, the primary missions of the air force were to "organize, train, and equip for prompt and sustained combat operations in the air, gain and maintain general air supremacy, defeat enemy air forces, and control vital air areas."[33] Later in the early 1950s, two air force generals and future space advocates Homer Boushey and Thomas White cited these basic missions to establish a broad foundation to claim that air and space operations blended into one operational medium. Back in the late 1940s, however, the leadership of USAF did not advocate this broad interpretation of the function of the air force. Rather, USAF leaders such as LeMay, who became the commander of Strategic Air Command in 1948, focused on building a robust, manned, strategic bomber force rather than a space force. LeMay's decision to focus on building strategic bombers stifled the air force support of satellites.

Despite the mainline trend, some air force officers saw air and space as one continuous operational medium. For this small cadre of space-minded officers, the additional functions of USAF outlined by the Joint Chiefs in 1948 offered more narrowly defined missions for the air force to establish its future claim as a space force. Although some specified functions of the air force had no obvious implications for space policy, several did. For example, the Joint Chiefs of Staff defined USAF responsibility for "strategic air warfare, development of equipment, doctrine, and procedures for air defense of the United States, and the furnishing of aerial intelligence."[34] Still later in the 1950s, USAF would argue that these responsibilities applied to satellites and space systems.

Several of these defined functions paralleled the potential use of satellites as identified by Ridenour in 1946. Specifically, Ridenour believed, as already mentioned, that satellites could provide "aerial photography, tactical and strategic reconnaissance, cartography, communications, and collection of weather data." By the end of the 1940s, the air force worked feverishly at conducting high-altitude photoreconnaissance to collect data on the Soviet Union and its allies.[35] With the Soviet Union's first atomic bomb test and growing concerns of a bomber gap, Truman's concern over the loss of the US monopoly in atomic weapons reinforced the need to monitor the strategic forces of the Soviet Union by gathering intelligence. The air force conducted the bulk of missions designed to collect photographic, electronic, and signals intelligence on the Soviet Union. Although Ridenour identified satellites as potentially useful systems in these missions, the air force remained fixated on building a strategic air force, rather than developing satellites as intelligence-gathering platforms.[36]

The budget and demands of the security environment dictated that the United States armed forces deemphasize the development of satellite and missile technology in the late 1940s, which left satellites and space programs to be thought about by a small, yet dedicated group of people. Throughout the remainder of Truman's tenure as president, RAND Corporation employees Louis Ridenour, James Lipp, and Paul Kecskemeti worked with air force officers Bernard Schriever and C. P. Cabell in the development of satellites for military use. This group continually strove to define the practical use of satellites. Later USAF used their groundbreaking work to claim space operations as simply an extension of its atmospheric missions.

Despite the work of the small group of RAND employees and USAF officers, the Truman administration in the second half of the 1940s kept the Department of Defense budgets small for research and development of missiles and rockets by the army, navy, and air force.[37] In 1947, the air force lost funding for its ballistic missile programs, while the navy and army continued work on their respective Viking and Redstone missiles.[38] During 1946–1950, the air force, despite a cautious and steady approach to the use of satellites, remained focused on devoting a large portion of its budget to developing a robust strategic bomber force.[39] Even though LeMay had been instrumental in initiating USAF interest in satellites, by 1948 his priorities had changed. As commander in chief of Strategic Air Command (CINCSAC), LeMay focused his attention on strengthening the US manned strategic bomber force.[40] The bitter fight between the navy and the air force over the development of the B-36 and the B-47 and follow-on B-52 bombers required LeMay to guard the development of SAC vigilantly. Partly because

(Left to right) Dr. Hugh L. Dryden, Dr. Ben Lockspeiser, Dr. Theodore von Karman, and Dr. A. P. Rowe tour aeronautical research facilities in Germany, May 9, 1945. NASA, courtesy of nasaimages.org

of LeMay's efforts, US strategic war plans between 1947 and 1952 heavily emphasized manned strategic bombing.[41] LeMay's emphasis on building a robust strategic bomber force and developing war plans reduced the likelihood that the air force would spend scarce defense dollars on satellites, but USAF continued to fund the RAND Corporation's research on the use of satellites for military purposes.

The central role played by the RAND Corporation in promoting America's use of satellites for military applications did not first develop because of the demands of the Cold War arms race. Rather, RAND's involvement with satellites blossomed immediately after the close of World War II because AAF general H. H. Arnold wanted to maintain a close relationship between universities and military scientists, and because LeMay wanted work to proceed beyond the research completed by the navy. Arnold believed that this relationship had been pivotal in the nation's success in World War II and needed to be sustained. To reinforce the relationship between the military and scientists, the air force established its own Science Advisory Board (SAB) as well as the Air Research and Development Command (ARDC) to conduct vital aeronautical (and later aerospace) research and development.

These early actions provided the US government with a foundation to begin building an interest in the military uses of space.

The Air Force and Satellites

In the late 1940s, the air force claimed space as its domain. Although the air force as a whole lacked a doctrine or even a strategic concept for using space, LeMay and Schriever advanced the view that air and space were actually an integrated whole—"aerospace"—a decade before the word came into use in the air force and became official doctrine.[42]

The foundational generalizations in RAND's feasibility report far exceeded the scope and detail developed in the navy's initial feasibility study on satellites in 1945. Beyond just analyzing the technological feasibility of satellites, RAND's engineers outlined their potential military and civilian uses. While the air force generally, and LeMay in particular, focused mostly on other strategic issues for the remainder of the 1940s, the staff at RAND and Bernard Schriever and his ARDC staff worked throughout the last years of the decade justifying the military use of satellites.

Identifying the likely military and civilian uses of satellites in the early days of the space age, RAND's staff recognized that satellite development benefited the scientific and military communities but cautioned: "Attempting to estimate the value to be derived from a development program aimed at the establishment of a satellite circling the earth above the atmosphere is as difficult as it would have been some years before the Wright Brothers flew at Kitty Hawk to visualize the current use of aviation in war and peace."[43] Aware of the obstacles, but impressed by the possibilities of satellites, a cadre of air force officers supported their use for military purposes. Simultaneously, RAND worked closely with the air force to accept satellites as a key to its future. The Truman administration did not share the vision, and his assistant secretary of war for air (later the first secretary of the air force) Stuart Symington did not see any reasonable prospect for space as a military environment. Despite the administration's reluctance to support the development of satellites, the air force and RAND built the theoretical foundation to use space for military and scientific purposes.

On May 13, 1946, the same month in which RAND published its first satellite feasibility study, Lt. Gen. Ira C. Eaker, deputy commander of Army Air Forces, sent a memorandum to Assistant Secretary of War Symington. Eaker described a proposed scientific military experiment to attempt to launch a projectile to the moon using a captured V-2 rocket. Eaker viewed this proposed experiment as "an entry into a completely unknown field," adding that "it is difficult to anticipate what scientific discoveries might result from this experiment." Eaker held that the lunar probe "opened new avenues of scientific endeavor" to the United States.[44] Specifically, he believed that a lunar probe could provide data on the outer ionosphere and the extreme temperatures in outer space.

Finally, Eaker argued: "If we may assume that the future of air conquest will bring with it a conquering of outer space, then clearly this experience and the enthusiasm which this project might generate will be very beneficial in the long run."[45] Despite Eaker's support for the project, Symington simply responded: "After reading this it looks to me as if we will be sticking our neck out" in a "careless use of taxpayer's money." Symington's response to the moon experiment typified the Truman administration's reaction to space experiments. Concerned over the need to reduce the defense budget from the very high levels of the wake of World War II, Truman and the other ranking members of his administration, such as Symington, put the broader strategic needs of the nation over funding research and development of space experiments and satellites.[46] Because of other Cold War strategic concerns, Truman effectively left the air force and RAND Corporation to focus on the theoretical utility of satellites. While the navy canceled its work on satellites due to lack of funds in 1946 and 1947, the air force maintained its interest by authorizing further feasibility studies and research on the military uses of satellites.

The Feasibility of Satellites

With support from the air force, the RAND Corporation between 1949 and 1952 continued to modify the prospective military applications of satellites by holding internal conferences and writing research reports. According to air force historian Lee Bowen, Air Materiel Command, prompted by RAND's work in December of 1947, "recommended the establishment of an Air Force satellite program."[47] To advance the proposed satellite program, in January 1948, the air force deputy chief of staff for materiel, Lt. Gen. H. A. Craig, declared "that with the passage of time" a satellite vehicle would become fiscally feasible.[48] Craig forwarded the satellite program up the chain of command by recommending to the USAF vice chief of staff, Gen. Hoyt S. Vandenberg, that the air force pursue a satellite program. Persuaded by Craig, Vandenberg issued a supportive policy statement, stating: "The USAF, as the service dealing primarily with air weapons—especially strategic—has logical responsibility for the satellite. Research and development will be pursued as rapidly as progress in the guided missile art justifies and requirements dictate. To this end, the program will be continually studied with a view to keeping an optimum design abreast of the art, to determine the military worth of the vehicle."[49] With Vandenberg's announcement, the air force tentatively embraced satellites. Though Vandenberg stopped well short of recommending the building of satellites immediately, he asserted the air force claim to satellites as part of the USAF mission. Vandenberg thought satellite systems would be feasible only after the US missile program matured. By the late 1940s, the air force had officially recognized satellites as a technical possibility with military and scientific applications.

Lacking an adequate rocket booster, air force officers Schriever, Craig, and Vandenberg, with the assistance of Louis Ridenour, Amron Katz, and Paul Kecskemeti of the RAND Corporation, worked at defining the specific military missions for satellites. To consider the potential military use of satellites, the RAND Corporation hosted an internal three-day conference in January 1949 entitled "On Methods for Studying the Psychological Effects of Unconventional Weapons," to discuss and debate the psychological impact of "atomic weapons, radiological weapons, economic weapons, psychological warfare, and satellite vehicles."[50] On January 28, 1949, the last day of the conference, the participants discussed the "capabilities and utility" of satellites.[51]

From the onset of the discussion, the head of the RAND Corporation's Missile Division and chair of the conference, James Lipp, declared satellites to be "qualitatively different from other weapons, as their primary purpose is not to destroy things." Lipp outlined missions such as "communications, observation, and scientific measurements" to reinforce his characterization of satellites as non-aggressive. Lipp restated the basic missions identified by Ridenour in 1946. Like Ridenour, the conference participants rejected the idea of using satellites as exo-atmospheric bombing platforms because the size and weight of atomic weapons were too great and the rocket boosters of the era lacked the sufficient power to lift a bombardment satellite.

After RAND personnel discussed the basic character and use of satellites in their January 1949 conference, the participants focused on the prestige the United States could garner from the public launch of a satellite. RAND engineers W. P. Davidson, A. Kaplan, and Lipp voiced their beliefs that a US. satellite would serve to demonstrate the advanced state of the scientific and technological capabilities of the United States to the world.[52] The conference participants believed satellites could significantly influence international opinion about US advanced scientific and technological capabilities.

Ultimately the conference discussions focused on six basic themes: "evidence of advanced technology, psychological uses, special payloads, continual reconnaissance, public announcement of the program, and proper launch times."[53] Guided by their thematic discussions, RAND employees established a broad understanding of how the United States could use satellites as unconventional psychological tools to influence the international community. The ideas and issues identified by RAND's engineers, social scientists, and strategists in this conference established the basic parameters upon which Presidents Eisenhower, Kennedy, and Johnson would forge a national space policy designed to militarize space, while also maintaining space as a weapons-free frontier for all nations to explore. Conference chairman James Lipp spearheaded the development of these theoretical ideas that would serve as the future tenets of US space policy.

Specifically, Lipp reasoned, for example, that all nations should "cooperate in the development of space navigation" because of its "universal significance."[54] Furthermore he reasoned that international cooperation would be more difficult if the space environment was complicated with active weapons.

Having outlined the use and psychological impact satellites could have on the international community, the RAND personnel moved beyond their initial technical feasibility studies and conference discussions to focus on the military missions of satellites and their ability to contribute to the national security goals of the United States.

In the remaining years of the Truman administration, the RAND Corporation continued to advance and refine its work on the utility of satellites. Under contract from the air force, RAND social scientist Paul Kecskemeti analyzed the political usefulness of satellites with the publication of "The Satellite Rocket Vehicle: Political and Psychological Problems" in October 1950.[55] Kecskemeti summarized that satellites served US national security and political interests by demonstrating the advanced scientific and technological capabilities of the United States to the international community. He claimed that the use of "novel and unconventional" systems would provide strategic and political benefits to the United States. In essence, Kecskemeti argued that the United States could use satellites to improve its international prestige, which he identified as the primary "political" mission of satellites. Beyond boosting US prestige, Kecskemeti reasoned that satellites provided "better intelligence about Soviet installations." Kecskemeti's assessment and theory echoed Truman's concern over the need for strategic intelligence about the Soviet Union, as well as maintaining America's technological and scientific prestige. In the aftermath of the Soviet Union's surprise test of its first atomic bomb, these two issues struck a chord with the Truman administration.[56]

Beyond Kecskemeti's analysis of a satellite's ability to provide strategic intelligence, he reasoned that the United States needed to exploit satellites for prestige purposes.[57] For Kecskemeti, the satellites could improve the position of the United States vis-à-vis the Soviet Union.[58] Kecskemeti spent the majority of his report considering the impact of America's use of satellites on national prestige, rather than analyzing the intelligence-gathering capabilities of satellites.

In his analysis, Kecskemeti argued that satellites could have a political impact on allies, neutrals, and hostile nations by showcasing America's "superior capabilities."[59] Kecskemeti reasoned that satellites would reassure US allies and neutral nations of America's superior technology. In considering hostile nations, Kecskemeti believed that a satellite program stood as a beacon of US military power by demonstrating the superior scientific and technological capabilities. Kecskemeti held that convincing the Soviet Union of America's "overwhelming strength" remained a main objective of satellites.

Kecskemeti's work in 1950 extended and refined the basic ideas identified by RAND between 1946 and 1949. Deemphasizing the satellite's ability to gather intelligence, Kecskemeti stressed the importance of international prestige. Judging from Kecskemeti's analysis, the political and psychological role of satellites in the first decade of the Cold War remained equally important as intelligence gathering. Throughout the Cold War, US and Soviet leaders continually boasted that their space programs (military as well as civilian) demonstrated national strength as well as advanced scientific, technological, and engineering capabilities. The leaders of both nations considered these programs imperative in the Cold War struggle.

Whereas Kecskemeti favored the political mission of satellites, USAF and RAND worked jointly on the practical and strategic uses of satellites. On March 17, 1951, the USAF director of intelligence, Gen. Charles P. Cabell, sent a memorandum to Colonel Schriever outlining the technical requirements and intelligence goals for the development of a USAF satellite.[60] Cabell emphasized that the satellite needed to have the photographic capability to gather intelligence on "harbors, airfields, oil storage, large residential areas, and industrial areas."[61] Cabell also insisted that the satellite have specific capabilities to "cover the entire USSR in a period of weeks and furnish continuous daytime observation." Cabell wanted the satellite to produce a photographic record of the Soviet Union that the air force could use for "radar scope navigation, bombing, revision of aeronautical charts, and meteorological study." Cabell stressed that the United States "urgently needed a reconnaissance system" that could provide vital intelligence.[62]

The requirements identified by Cabell prompted USAF to ask RAND to continue researching the practical application of satellites for military missions. In April 1951, RAND employees S. M. Greenfield and W. W. Kellogg completed a study, "Inquiry into the Feasibility of Weather Reconnaissance from a Satellite Vehicle," which supported the USAF claim that satellite systems capable of providing data for "meteorological study" were needed.[63]

Greenfield and Kellogg moved beyond the psychological and political use of satellites that predominated in RAND's earlier satellite studies. Instead they focused on the practical applications and operational consequences. In the opening of their report, for example, Greenfield and Kellogg reasoned: "[I]n the event of armed conflict, aerial weather reconnaissance over enemy territory will be difficult to obtain."[64] Unlike the experience in World War II, when the air force flew over enemy territory to collect weather data, the technology and weapons that would be used in a future war made such manned overflights extremely risky. Satellites offered an alternative to pilots flying weather-data missions.

Greenfield and Kellogg stated that the use of a satellite could be done with a "high degree of safety and a reasonable probability of success." Using the advantages of high altitude and advanced television sensors, satellites, according to Greenfield and Kellogg, could become "un-paralleled instruments for weather reconnaissance." Because US strategic war plans relied heavily on strategic bombers, USAF needed to have weather data for potential target areas. Although the military services of the United States continued to use aircraft and balloons to gather weather data, satellites possessed advanced capabilities beyond traditional forms of data collection. The need for constant and updated information drove the air force and other services to embrace the use of satellites for weather reconnaissance.[65]

Promoting the use of satellites as weather platforms reflected the belief that satellites were non-aggressive military tools. In the coming decades, this distinction became a central element in the construction of the US military space program and the national space program. Throughout the remainder of Truman's tenure, the idea appeared infrequently in the administration's foreign policy goals.

Psychological Strategy Board member George Morgan characterized the Truman administration's foreign policy goals in May 1952 with broad rhetoric, which paved the way for future presidents to justify the future US national space program in an altruistic manner. Morgan stated that the foreign policy of the Truman administration focused on "the defense of the United States, creation and maintenance of the structure of world peace in accordance with the Charter of the United Nations, and the development of conditions in which peoples may freely establish the governments and institutions under which they live."[66] Within these broad objectives, Morgan stressed that "armament is a temporary shield necessary to hold an aggressor in check." The need for the United States to have good intelligence and be aware of the Soviet Union's advances in strategic and tactical military capabilities would lead Presidents Eisenhower, Kennedy, and Johnson to justify US military satellites as vital to the foreign policy and national security objectives of the United States as well as critical to the maintenance of world peace.[67]

The work done by RAND's employees proved imperative in America's movement to militarize space and secure space as a weapons-free frontier. Although the United States was still several years away from an operational satellite, the process of defining and justifying this new technology for military purposes evolved from 1946 to 1952.

The feasibility studies and conferences conducted at RAND, with cooperation from the air force during those years, provided the engineers, strategists, social scientists, and officers involved in satellite research with a foundation to transform satellites from a theoretical possibility into a nonaggressive conceptual military system. The evolution in thinking concerning the usefulness of satellites for military missions followed a pattern from 1946 to 1952. First, the navy and air force worked with North American Aviation and the Douglas Aircraft Company in identifying the technological feasibility of satellites. After confirming the feasibility of satellites, the air force granted the RAND Corporation contracts to study the psychological and political value of these new "unconventional" systems. After studying and defining the psychological and political potential of satellites, RAND's Kecskemeti, Greenfield, Kellogg, with Cabell and Schriever in the air force, defined the operational use of satellites. Within this ongoing technological evolution, these men identified satellites as something inherently different from weapons, because they collected data and gathered intelligence as opposed to killing or destroying targets as their primary objective.[68] Despite the volume of work and theories advanced about the uses of satellites in the initial stage of the space age, the idea of using satellites as aggressive weapons received minimal attention.

The research done by RAND employees and the air force throughout the tenure of President Truman established the foundation for the military's use of satellites for non-aggressive purposes. While a minority opinion evolved advocating the weaponization of space in the 1950s, future Presidents Eisenhower, Kennedy, and Johnson would remain committed to the goal of using satellites for non-aggressive military missions and strive to keep space free of weapons. The rudimentary work completed on satellites in the Truman era provided the guidance for the future development of a robust military space program and the eventual treaty banning weapons of mass destruction from space.[69]

Conclusion

Although often overlooked historically, but vital in understanding the developing of US national space policy, a small cadre of RAND personnel and USAF officers did research during Truman's presidency concerning the use of satellites for military purposes. While Truman did not offer his outright support for the research and development of satellites during his tenure as president, others did. USAF officers Bernard Schriever, C. P. Cabell, and Hoyt Vandenberg, and employees of the RAND Corporation such as S. M. Greenfield, James Lipp, and Paul Kecskemeti defined how satellites could be used for political and military purposes.

Within this initial stage of the space age, the idea raised by James Lipp in 1949 that satellites were something inherently different from weapons systems emerged as an important step in maintaining space as a weapons-free frontier. Future presidents Eisenhower, Kennedy, and Johnson would recognize the "universal significance" of space and openly support US military satellites and space programs as essential to the national security of the United States and the preservation of world peace. Truman's successors would use these ideas to justify America's commitment to its military and civilian space programs and the peaceful use of space. Furthermore, Eisenhower, Kennedy, and Johnson would use the US space program to advance a pacific image of the United States while highlighting the Soviet Union's perceived interest in dominating space. During the Cold War, the United States perceived that the Soviet Union desired to weaponize and control space. This represented the antithesis of the American ideal.

With the research done by RAND, the air force staked an early claim as the nation's space force by identifying space as an extension of airpower missions. In the 1940s, this position, a forerunner of the concept of "aerospace," provided the Air Force with a claim to be the nation's space force.

The initial decision by General LeMay in 1946 to authorize RAND to present the air force with a feasibility study on satellites proved pivotal in later years. The rivalry between the navy and air force in 1945–1946 over satellites prompted the air force to begin a small, yet determined program to outline the military, psychological, and political usefulness of satellites. Despite the decision to begin looking toward space as a medium for military operations, the air force remained deeply committed to enhancing its strategic bomber force before venturing into space. Yet fundamental work done by RAND and key USAF personnel between 1946 and 1952 provided the United States with a conceptual and operational foundation to move toward militarizing space, while also seeking to keep space as a weapons-free frontier. By the time of Eisenhower's inauguration in 1953, the space age had already begun and the tenets of America's future national space policy had been defined.

Chapter 2: Embracing the Militarization of Space, 1953–1960

Military space activity dramatically increased during Dwight D. Eisenhower's administration. Unlike his predecessor, Eisenhower took an active and direct interest in space-related issues and, more specifically, he firmly embraced the use of satellites for military and peaceful missions. Building upon the research done on satellites by the RAND Corporation and the air force during Truman's tenure, Eisenhower codified these efforts with the establishment of a national space policy.

Moreover, while Eisenhower presided over the militarization of space he also lobbied the United Nations for an international resolution to preserve space as a weapons-free frontier for all nations to explore. Eisenhower faced increased pressure from the military to protect the emerging extraterrestrial interests of the United States, which the military services believed to be of vital geostrategic importance in the Cold War contest with the Soviet Union. To appease the services, Eisenhower authorized preliminary research on antisatellite systems (ASAT) and ballistic missile defenses (BMD). Eisenhower supported these programs because he believed that they could be used to stop a Soviet attempt to control space. Though he was against the use of weapons in space, Eisenhower defended US interest in ASAT and BMD research as necessary to preserve global peace and stability. Although Eisenhower supported R&D on ASAT and BMD systems, he stopped short of authorizing deployment. While ASAT and BMD research began in the Eisenhower years, they blossomed into fully funded programs under the direction of Presidents Kennedy and Johnson.

Eisenhower did not usher America into the space age. It was already there, and he expanded on principles set out in the Truman years. What was new in the Eisenhower presidency was the president's recognition of the need for a national space policy based upon the militarization of space, while also working toward an international agreement to ban weapons from space.

President Eisenhower gets a briefing from Dr. Wernher von Braun on the Saturn 1 rocket at the Marshall Space Flight Center, September 8, 1960. NASA, courtesy of nasaimages.org

The Cold War Security Environment and Space

Eisenhower's space policy evolved within the competitive parameters of the Cold War; therefore, the bipolar rivalry between the United States and the Soviet Union was an essential variable in the development of America's early space programs and policies. In documents from the National Security Council (NSC), the tense rivalry of the Cold War is evident. The security concerns advanced by the Technology Capabilities Panel (TCP) of the president's Science Advisory Committee and headed by MIT scientist James Killian, as well as in NSC meetings, resulted in the formulation of policy papers such as NSC-162/2 and NSC-5520 that justified the use of space for national security purposes.

Although the NSC produced NSC-68 during Truman's presidency, Eisenhower relied on its comprehensive recommendations to strengthen the military forces of the United States. Even though NSC-68 does not directly discuss space programs, Eisenhower used its basic foundations to frame a political argument for the militarization of space. NSC-68 specifically suggested that the United States needed to maintain a strong international presence to frustrate the perceived designs of the Kremlin.[1] Armed with NSC-68, Eisenhower emphasized a restructuring of the US military to counter the expanding international influence of the Soviet Union and the threat posed by the exportation of communism. He believed that satellites provided the United States with the necessary intelligence-gathering capabilities to break the Soviet Union's shroud of secrecy. For Eisenhower, satellites and space systems played vital roles in the national security strategy of the United States.

The first steps toward the development of Eisenhower's space policy began in October 1953. In NSC-162/2, Eisenhower's National Security Council formally recognized the Soviet Union as the primary threat to the United States.[2] The authors of NSC-162/2 identified the maturing atomic capabilities of the Soviet Union as well as the international exportation of communist ideology as primary threats to the security of the United States. To counter these threats, the creators of NSC-162/2 advocated a "strong military posture" and the "maintenance of morale and free institutions" as hedges against the Soviet Union's advance. Specifically, the NSC advised Eisenhower to expand the scientific and technical training of US citizens in an effort to remain the "leader in the free world."[3] In the 1950s, science and technology served as national indicators of strength and the NSC believed the president needed to maintain and expand the international prestige of the United States.

The need for the United States to remain the leader of the free world became a constant refrain under Eisenhower, even if the idea was not new. In 1945, H. H. Arnold and aeronautical scientist Theodore von Karman had heralded the importance of maintaining a strong aeronautical research and development program. Two years later, Truman's Air Policy Commission urged him to develop a robust international image for the United States by investing in advanced aeronautical science and technological research. Eisenhower and his administration continued to advocate the international importance of technological superiority. So, even though NSC-162/2 did not focus on the use of space or the need to develop a space policy, the National Security Council did stress the need for effective intelligence gathered by advanced means, which raised the possibility of using space-based intelligence-gathering systems. The emphasis on science and technology dovetailed with the intelligence requirements stipulated in NSC-162/2. The bipolar rivalry of the Cold War and America's desire to retain its position of scientific and technological superiority led the Eisenhower administration to define requirements for US space programs and policies in 1955.

At the direction of Robert Lovett, Truman's last secretary of defense, A. Y. Grosse, the director of Temple University's Research Institute, led a study of US satellite research. Lovett had commissioned the report in February 1952 to grasp the issues involved in building a satellite, but did not finish the report until spring 1953, the start of the Eisenhower presidency.

Grosse recounted the "scientific, military, and psychological value of an unmanned satellite."[4] Reinforcing the research already conducted on the use of satellites from 1946 to 1952, Grosse stressed that satellites could "obtain valuable scientific information regarding the conditions in outer space," as well as serve as "valuable observation posts" for the military, and provide an effective method of "broadcasting messages to the free world." Grosse did little more than reiterate the ideas established earlier by members of the RAND Corporation and of the armed services, especially USAF, but his study highlighted the issues for the new president. Grosse discussed the scientific and military applications of satellites, but he concluded that the psychological effect of satellites would be "of the utmost value" to the United States.

Concerning the psychological benefits, Grosse maintained that "a satellite would have the enormous advantage of influencing the minds of millions of people the world over during the so-called Cold War." Considering the technological rivalry between the United States and the Soviet Union, Grosse cautioned: "Since the Soviet Union has been following us in the atomic and hydrogen bomb development, it should not be excluded that the politburo might like to take the *lead* [his emphasis] in the development of a satellite. They may also decide to dispense with a lot of the complicated instruments that we consider necessary to put into our satellite to accomplish the main purpose, namely of putting a visible satellite into the heavens first. If the Soviet Union should accomplish this ahead of us it would be a serious blow to the technical and engineering prestige of America the world over."[5] Grosse's strong advocacy of satellites to produce the desired psychological effects supported Eisenhower's belief that the United States needed to maintain and enhance its international prestige vis-à-vis the Soviet Union. Beyond the use of satellites for bolstering prestige, Eisenhower recognized their ability to gather and collect reconnaissance and intelligence on the Soviet Union. Acknowledging the need to avoid a "nuclear Pearl Harbor," Eisenhower established his own panel to consider the application of advanced technologies to national security issues.

Eisenhower identified satellites as valuable national security assets based on the recommendations of the Technological Capabilities Panel (TCP). The TCP worked to define how the United States could use science, technology, and engineering to lessen the president's concerns over the Soviet Union's strength. In February 1955, the members of the TCP drafted their report entitled *Meeting the Threat of Surprise Attack.* The members of the panel examined the vulnerability of the United States and assessed ways in which science and technology could be applied to rectify this "hazard."[6] Members of the TCP stressed that "technical innovations could be powerful instruments for creating strength," as well as providing a "deterrent to war." The TCP's members identified intelligence and communication as two areas where the United States could exploit the use of satellites. Against the administration's increased concerns about the Soviet Union's expanding military threat, the TCP's members emphasized the immediate need to expand America's nonhuman intelligence capabilities. The members' recognition of the need to use advanced technological means to collect military intelligence on the Soviet Union led them to envision the use of satellites as an advanced form of gathering intelligence.

In 1955, the United States already used signals intelligence (SIGINT) equipment to collect vital military, political, and diplomatic data, but the nation needed additional information on the Soviet Union's strategic forces. The members of Killian's TCP also stressed that "revolutionary new techniques needed to be devised to give the United States facts and answers instead of assumptions and estimates."[7] With the Air Policy Commission's serious miscalculation of when the Soviet Union would first test an atomic bomb still fresh in the collective memory of the administration, Eisenhower wanted improved US intelligence gathering and reconnaissance capabilities over the Soviet Union. The TCP's members responded to the president's wishes by recommending the use of satellites for reconnaissance.[8]

The TCP advocated the construction of civilian research satellites as a way to allow the technology for military reconnaissance and intelligence-gathering satellites to mature. The members of the TCP reasoned that the civilian satellite program would yield data and techniques to ease the construction of military programs such as advanced intercontinental ballistic missiles (ICBMs), SIGINT gathering platforms, photoreconnaissance, and secure overseas communication capabilities.

The TCP members' belief that satellites could be a useful military medium received the highest support within the administration. Eisenhower accepted the urgency and the paramount need for the United States to keep track of the Soviet Union's growing nuclear arsenal, a concern which had been advocated nearly a decade earlier by H. H. Arnold and Theodore von Karman in 1945. Borrowing from Arnold and von Karman, the TCP again stressed the strategic importance of investing in scientific and technological research as vital to America's national security.

Eisenhower and his administration, most notably Secretary of State John Foster Dulles and Director of Central Intelligence (DCI) Allen Dulles, became vehemently concerned about the Soviet Union's capability to launch a surprise attack on the United States. With the DCI's growing concerns over the Soviet Union's development of ICBMs, satellites emerged as vital components in the US quest for security and technological superiority. These issues facing Eisenhower and his administration paralleled the issues and ideas debated by the RAND Corporation and USAF in the Truman era when satellites were first identified as possessing military potential. The major difference between the two eras was evident when Eisenhower directed his NSC staff to define a space policy and to move forward with the construction of satellites.

Space Policy prior to *Sputnik*

By the middle of the 1950s, Eisenhower firmly embraced satellites as useful for intelligence gathering, as well as for bolstering the international prestige of the United States.[9] Eisenhower's support for the use of satellites for military purposes led to America's first official space policy. With his approval, the NSC issued NSC-5520, U.S. Scientific Satellite Program, in May 1955. Based upon the intelligence and security needs identified by the TCP, the National Security Council stipulated in NSC-5520 that the "United States was believed to have the technical capability to establish a satellite program in the near future (1957–1958)."[10] The NSC members also highlighted that on April 16, 1955, the Soviet Union had announced their creation of a "permanent high-level, inter-departmental commission for interplanetary communication."[11] Following the ideas advanced by the TCP, which paralleled those raised in 1946 by members of Project RAND, the National Security Council in NSC-5520 directly stated that "considerable prestige and psychological benefits will accrue to the first nation which is successful in launching a satellite." The "advanced technology" associated with a successful satellite program represented not only scientific and technological superiority for Eisenhower and his administration, but also the advanced military capabilities of the United States. The authors of NSC-5520 predicted that satellites "might have repercussions on the political determination of the free world to resist the Communist threat." These ideas were not original to Eisenhower and his staff. The policy makers of the Eisenhower era simply restated and made official the beliefs and ideas about the military application of satellites already well understood in USAF and by the RAND Corporation.

Recognizing that satellites could influence international opinion, Eisenhower remained determined to present a strong, yet pacific image as to America's interest into the uses of space. To counter any international suspicion that the United States sought to control space, the National Security Council identified the International Geophysical Year (IGY) from July 1957 to December 1958 as an opportunity to demonstrate America's commitment to space as an open scientific frontier where all nations would benefit. Parallel to NSC-5520, Eisenhower staffers believed the IGY was an opportunity for the United States "simultaneously to exploit the over-all benefits" of the scientific satellite program (Project Vanguard), while also "capitalizing on the research" to benefit the military satellite program.[12]

The satellite program advocated in NSC-5520 demonstrated that the Eisenhower administration built upon the accepted ideas of the Truman era and recognized the consequences of losing to the Soviet Union in the race to launch the first man-made satellite. Eisenhower strove to boost the scientific and technological prestige of the United States, while also improving America's reconnaissance and intelligence-gathering capabilities over the Soviet Union.

The Vanguard satellite program received a huge boost from NSC-5520. Using NSC-5520, the members of the NSC staff recommended that the Department of Defense build "six instrumented satellites to be used within the IGY, with the expectation that one would orbit [the earth]."Earth.[13] The NSC recommended the development of an additional six satellites for the IGY. Considering the proposed twelve satellites through the spring and summer 1956, the continual refrain from NSC members was the belief that America needed to build civilian satellites as a way to demonstrate its commitment to the use of space for peaceful purposes.[14]

Furthermore, the NSC staffers recognized that the Vanguard program afforded a vehicle for launching military satellites, provided security could be maintained.[15] The need for military satellites as well as the need to convince the international community of America's scientific and technological superiority dovetailed in NSC-5520. Interestingly, while Eisenhower publicly touted the scientific and civilian orientation of the Vanguard satellite program, the Department of Defense (DoD) carried out the majority of work.

Eisenhower understood that this research could be translated into a future military satellite system. Therefore despite his public rhetoric over the separation between the civilian and military space programs, Eisenhower understood the overlapping nature of satellite development as a critical element in the development of a national space policy.

In October 1956, DoD issued its *Progress Report on the U.S. Scientific Satellite Program* with the prediction that the first launch date for a US satellite would be no earlier than October 31, 1957. By the date of this report, the satellite program had incurred substantial cost overruns and engineering obstacles. The increase in the cost of the program, from an initial figure of twenty million dollars to approximately sixty-three million dollars, led the authors of the report to suggest that the program be scaled back to six satellites.[16] As the Vanguard satellite program evolved through 1956, Eisenhower administration officials, such as Special Assistant to the President Nelson Rockefeller, Deputy Secretary of State Christian Herter, Secretary of State John Foster Dulles, and Director of Central Intelligence Allen Dulles, firmly believed the United States stood to benefit from the prestige and psychological effect of being the first nation to launch and orbit a satellite. In 1956, Eisenhower publicly emphasized the importance of the Vanguard satellite program to the IGY over the military value of intelligence-gathering satellites.[17]

The Intelligence Advisory Committee of the Central Intelligence Agency (CIA) informed the president in November 1956 that the Soviet Union "would orbit a satellite any time after early 1957."[18] Concern rose in the administration over the Soviet Union's satellite program because the intelligence suggested that the Soviet Union had begun work on a satellite as early as 1954.[19] The administration's focus on the benefits of becoming the first nation to orbit a satellite led Allen Dulles to argue in January 1957 that the Soviet Union "was making a major effort to be the first nation to launch a satellite." Concerned about the Soviet Union's progress, Dulles provided the NSC with intelligence on the Soviet Union's recent test of a "one to five kiloton atomic weapon affixed to a missile." He suggested that what could launch a nuclear warhead might also be sufficient to launch a satellite. Dulles emphasized to the NSC the significance of this news with the reminder that the Soviet Union had conducted a similar test on February 2, 1956. Eisenhower faced the possibility that the Soviet Union would be the first nation to launch and orbit a satellite, as well as have the capability to launch a missile attack on the continental United States.[20]

After the DCI's briefing, Eisenhower queried the director of the National Science Foundation, Dr. Alan T. Waterman, as to the expected launch date of the first Vanguard satellite. Waterman told Eisenhower that he expected the launch date to be sometime in October 1957. Facing increased competition from the Soviet Union's satellite program, as well as a growing nuclear-tipped missile threat, Eisenhower and the NSC staff faced the possibility that the Soviet Union would garner increased prestige, as well as gain a strategic advantage.[21]

By May 1957, the Vanguard satellite program incurred additional costs as the program budget exceeded $110 million. The skyrocketing costs led the members of the NSC in the 322nd meeting of the National Security Council held on May 10, 1957, to debate the termination of the Vanguard satellite program.[22] During the meeting, Eisenhower proposed termination of the program because he believed that America should not race the Soviet Union, but rather build a space program to benefit the national security objectives of the United States. Moreover, he believed the funds could better serve in the development of useful military satellites to collect vital strategic intelligence, which the United States lacked.

America's need for intelligence and reconnaissance data on the Soviet Union's strategic and tactical capabilities influenced Eisenhower's thinking. In defense of the nonmilitary satellite program, Christian Herter and Allen Dulles argued that the termination of the program would provide the Soviet Union with a significant propaganda advantage that would ultimately affect America's international image. Through the course of the discussion, NSC staff members concurred that it would be best if the United States maintained the Vanguard program, despite the delays and increased cost, because the satellite still offered scientific and technological prestige.

Recognizing that the United States might not be the first nation into space, the US defense and scientific community emphasized America's qualitative superiority in electronic instrumentation.[23] Also, in an effort to depict the Soviet Union's satellite program as a military endeavor, Eisenhower emphasized the Vanguard satellite program as vital to the international cooperative effort of the IGY. Publicly embracing space as an open scientific frontier, Eisenhower championed the US-led IGY effort as a testament to the US

Vanguard satellite booster explodes on launch pad, December 6, 1957.
NASA, courtesy of nasaimages.org

commitment to work within the world community, while he also depicted the Soviet Union as a rogue nation bent on the military conquest of space.

Although the United States cared about increasing its prestige, the Soviet Union launched the first man-made satellite on October 4, 1957 and captured the prestige and psychological benefits of being the first nation in space. Eisenhower and his staff publicly de-emphasized the importance of *Sputnik,* although the Soviet Union's triumphant accomplishment seriously influenced the Cold War rivalry, and directly influenced the international community's perception of the scientific and technological capability of the United States.[24] Despite this public blow, Eisenhower remained committed to bolstering America's international prestige and improving its intelligence-gathering capability over the Soviet Union. Satellites served these functions for Eisenhower.

Military Space Programs prior to 1957

After the Soviet Union's launch of *Sputnik,* US citizens clamored for military space systems to protect the nation from the communist space threat. Although Eisenhower risked appearing out of touch with the American public, he did not advocate the need for space weapons. Rather he maintained his commitment to the militarization of space, by emphasizing the need for a reconnaissance satellite, a position that had been earlier recommended by the RAND Corporation in Project Feedback.[25] Despite the fact that USAF officers such as Chief of Staff Nathan Twining, SAC Commander Curtis LeMay, and Chairman of the Air Force Scientific Advisory Committee Jimmy Doolittle supported the project, concerns emerged in USAF over the schedules for developing the technology needed for operational satellite systems. The air force assigned the Feedback program to the newly created Air Research and Development Command (ARDC).[26] Within ARDC, the commander of the Western Development Division (WDD) of the air force, Col. Bernard Schriever, took charge of America's first military satellite program. The air force had to solve significant technical issues on the development of power units, boosters, and photographic equipment for the satellite before an operational system could be launched. Even though USAF embraced Project Feedback in 1951, it was not until 1953 that the recommendations of Feedback blossomed into a military program.

By 1955, Schriever and the personnel at WDD decided that the reconnaissance satellite program needed to be linked to the Atlas intercontinental ballistic missile (ICBM) program.[27] Schriever and his WDD staff, while working on the satellite program, also had the primary task of perfecting and building ICBMs. For the United States, both programs proved pivotal in the militarization of space, as ICBMs proved valuable as rocket boosters for satellites.

The work done by WDD on the reconnaissance satellite by 1953 had been given two designations. Schriever and his staff worked on the components of the system identified as "Project 409–09: Satellite Component Study." Schriever and the WDD designated the entire program "Weapon System 117 L" (WS-117L). Schriever and the WDD simultaneously attacked the technical obstacles associated with using a space-based reconnaissance platform and the work on ICBMs such as the Atlas and Titan. Both Atlas and Titan served as military satellite boosters as well as ICBMs.

With the work going forward on the technical side of the satellite reconnaissance system, LeMay and his Strategic Air Command requested the development of an intelligence platform to provide "prompt bomb damage assessment" (BDA) in a post-strike environment.[28] The request advanced by LeMay and SAC evolved from a 1951 directive by C. P. Cabell, then director of air force intelligence.[29] LeMay, as commander of SAC, requested that the photoreconnaissance satellite be capable of providing SAC's intelligence officers and targeteers with data to conduct "real time" BDA within a nuclear war. SAC intelligence officers believed that the resolution of photographs required for conducting BDA "was not great, since large craters, not cities or military bases," were the focus of such missions.[30] LeMay's request synchronized nicely with the proposals by RAND members J. E. Lipp and R. M. Salter in their 1954 "Project Feedback Summary Report."[31] To reinforce the capabilities of a reconnaissance satellite using television to relay its data back to Earth, Lipp and Salter identified potential targets as: "Airfields of all sizes, and possibly indications of activities on airfields. Industrial concentration, isolated or within cities. Large plants and possibly some indication of types of plant. Harbors and facilities such as graving docks and large ships. Transportation, power and communication networks, including switching yards, bridges, canals, power lines, and perhaps activities in these fields. Urban areas, including the density of built-up areas. Large military installations, including military camps and explosive storage."[32] These targets caught the attention of USAF intelligence officers. Unintentionally they were a repeat of the target sets first identified by C. P. Cabell in 1951. Specifically, Cabell had identified "harbors, airfields, oil storage, large residential areas, and industrial areas" as intelligence targets.[33] In 1954, Lipp and Salter confirmed the capability of WS-117L to provide the intelligence envisioned by Cabell in 1951, as well as to meet the requirements outlined by LeMay and SAC in 1954.

Based on the capabilities identified by Lipp and Salter, Schriever and WDD received approval to develop the WS-117L system in July 1956, and program development on USAF reconnaissance satellite commenced in October 1956. The Lockheed Aircraft Corporation received the initial contract for development and testing of a satellite system. In addition to its work on satellites, USAF also envisioned other space-based systems evolving from the WS-117L program.

In the development of the WS-117L system, USAF and Lockheed engineers envisioned more than just a photoreconnaissance satellite. The plan envisioned the development of a "family of separate systems and subsystems" for the collection of intelligence data. "Subsystem G" of WS-117L was an "Infrared (IR) detection and surveillance" component, a space-based, ICBM, early-warning satellite.[34] This subsystem of WS-117L detected the exhaust plumes of ICBMs and even high-altitude aircraft and relayed the information back to US bases.[35] Joseph A. Knopow became the project manger of Subsystem G after USAF awarded the WS-117L contract to Lockheed in October 1956. Knopow, an electrical engineer who had experience working in the Operations Analysis Office of USAF Directorate of Operations, set forth to perfect the technology for detecting ICBMs from space. The USAF embraced the use of satellites for strategic early warning. Knopow had to bridge a wide gap between the scientific theory and actual technological development of a working IR detection satellite. The launch of *Sputnik* on October 4, 1957, reinforced the need for a space-based early-warning detection system and Subsystem G of WS-117, or the Missile Defense Alarm System (MIDAS), by 1958 became a priority for the USAF.

Given Eisenhower's concern about the Soviet Union's strategic forces, the launch of *Sputnik* heightened the sense of urgency in the United States. Yet the president's reaction in the immediate days after the launch did not reflect the public's heightened concern. Rather, Eisenhower and his press secretary, James Hagerty, de-emphasized the political and military implications of the Soviet Union's satellite launch in an effort to calm the public.[36]

In his first post-*Sputnik* press conference, on October 9, 1957, Hagerty distributed a summary concerning the Vanguard satellite program and then congratulated the Soviet Union's scientists on the successful launch of the world's first human-made satellite. Eisenhower then took to the podium to answer questions. In response to a question from an International News Service reporter regarding the concept of "racing" with the Soviet Union, Eisenhower allowed that "the Soviet Union gained a psychological advantage throughout the world, but in the scientific sense," he added, *Sputnik* remained a "minor achievement."[37]

Counter to the public's concerns over *Sputnik,* Eisenhower remained calm. In his press conference on October 9, he stressed that the launch of *Sputnik* "simply demonstrated the capability of a thrust mechanism of a certain power, accuracy, and reliability."[38] Despite Eisenhower's outward casualness, the Soviet Union's technological feat elevated US citizens' concerns over the international image of the United States and reinforced in the minds of air force leaders the need to develop military satellites and space programs.

In a move to sway international opinion toward the US position, Eisenhower refused to allow the Vanguard satellite program to use military boosters.[39] Eisenhower believed that the Vanguard satellite program highlighted America's commitment to the use of space for peaceful purposes. To demonstrate this, and, as a way to improve the international image of the United States, Eisenhower maintained the US commitment to separate civilian and military space programs.

In April 1958, Eisenhower lobbied for the establishment of the National Aeronautics and Space Administration (NASA) to control and direct the US civilian space program. America's civilian space programs became the domain of this new governmental administrative unit. The formation of NASA formalized Eisenhower's commitment to the principle that space was a new frontier to be used for peaceful purposes.

Juxtaposing it to the US position, Eisenhower identified *Sputnik* as a "technocratic triumph" of the Soviet Union's militarism. He strove to portray the United States as a peaceful nation in stark contrast to the militarism of the Soviet Union. Members of Eisenhower's National Security Council, specifically John Foster Dulles and the deputy administrator of NASA, Hugh Dryden, believed that the new US civilian space administration could act as a propaganda vehicle to reinforce other countries' view of the United States as truly committed to open access to space for the benefit of all nations.[40]

The members of NSC defined America's new outer space policy in NSC-5814/1 in August 1958 and outlined for the nation the scientific and military significance of exploring space.[41] First, the NSC stated that "more than any other imaginative concept, the mind of man is aroused by the thought of exploring the mysteries of outer space." Second, the NSC highlighted the prestige of space exploration by emphasizing that "the technical ability to explore outer space has deep psychological implications" for the nation. Next, the NSC believed the race for the "conquest of space" generated national competition between the Soviet Union and the United States as each nation strove to be identified as "the world's leader in science, military, industrial, and technological capabilities." Finally, the NSC suggested, "space exploration offered opportunities for international cooperation."[42]

In defining the significance of outer space, the NSC restated established ideas. Specifically, NSC 5814/1 centered on the importance of the scientific and psychological uses of space, as the discussion among RAND's employees had done in 1949. Eisenhower realized that the Soviet Union's launch of *Sputnik* provided him with an opportunity to commit the United States to the non-aggressive military and civilian use of space.[43]

While the members of the NSC in August of 1958 clearly emphasized the nonmilitary elements of space, they also provided guidance on the development of military satellites. Moving beyond the established reconnaissance systems already in development, the NSC specifically highlighted "weather observation, communication, electronic counter-measures, and navigation" as "planned or immediate" military operations for satellites.[44] Moreover, the NSC underscored the "future possibilities" of space, which included ideas for deploying weapons in space. In NSC- 5814/1, possible weapons in space were limited to:

- manned defensive outer space vehicles that might capture, destroy, or neutralize an enemy outer space vehicle;
- bombardment satellites (manned or unmanned). It was conceivable that, in the future, satellites carrying weapons ready for firing on signal might be used for attacking targets on Earth;
- manned lunar stations, such as military communication relay sites or reconnaissance stations. Conceivably, launching of missiles to the earth from lunar sites would be possible.[45]

The recognition by the NSC in 1958 of the potential to use weapons in space marked a significant deviation from the ideas of the Truman era, and also ran counter to the official position of Eisenhower.

The Aerospace Concept and Weapons in Space

In the immediate aftermath of *Sputnik,* while Eisenhower worked to quell the public's worries over America's declining scientific and technological status, the air force used the event as a catalyst to lobby for the weaponization of space and establish itself as the nation's space force. In November 1957, Air Force Chief of Staff Thomas D. White outlined the concept of aerospace, while stressing the strategic value of space. In an address before the National Press Club, White claimed that "total air power is the sum of a nation's aeronautical and astronautical [capabilities]."[46] To reinforce his point White boldly announced that "there is no division, per se, between air and space. Air and space are an indivisible field of operation." This announcement was not an original idea, but rather the culmination of a decade of thought within the air force about the military usefulness of space. Since 1946, the air force had believed that it had a claim to space missions. Under this new "aerospace" theory, White claimed outer space as an extension of air operations.[47]

To reinforce USAF's claim, White reminded the audience that the air force had been "penetrating space for several years, with manned aircraft." Using the X-2 and X-15 aircraft as examples, White emphasized the idea that air and space together made up one "indivisible field of operation." Building upon this idea, White claimed that USAF was the nation's legitimate space force.

Presenting the evolution of airpower as a model for aerospace power, White stated:

> "Originally, this third medium [air] was used to exploit reconnaissance and even as late as the beginning of World War I, the aircraft was used primarily for this purpose. Those who have the capability to control the air are in a position to exert control over the land and seas beneath."[48]

White used historical examples to claim legitimacy for the air force space mission by restating the age-old military axiom of the need to seize the "high ground." In fall 1957, the prevailing national security concern within the air force was the Russian attempt to "gain control of air and space." Building upon White's ideas, airpower advocates Alexander de Seversky and Gen. Homer Boushey, the deputy director for research and development, Headquarters, USAF, wrote in defense of the strategic importance of space.[49]

In January 1958, long-time airpower advocate de Seversky supported the aerospace concept, asserting, for example, that "airpower is space power."[50] De Seversky reasoned that the "endless air ocean" should be "controlled by a single homogenous force," namely, USAF. The views of White and de Seversky corresponded with those of Billy Mitchell during his quest in the interwar period for an independent air force. However, de Seversky and White did not advocate the creation of an independent space force. They believed the air force was the only service capable of handling space operations. The identification and recognition of the aerospace concept by the air force was an attempt to suppress and to delegitimize the bourgeoning space programs of the US Army and US Navy.[51]

Boushey, in January 1958, also lobbied Eisenhower and the DoD for air force preeminence in space in a speech before the Aero Club of Washington D.C. Boushey boldly stated that "the conquest of space is vital to US interest" and that "space superiority will become the primary factor in the assurance of world peace." Boushey, like both White and de Seversky, asserted that "space is a natural and necessary extension of our present air force operations."[52]

In the months after *Sputnik,* the air force restated its long-held position on space. Since the first air force feasibility study on satellites was begun in 1946, air force leaders such as Schriever, White, and Boushey recognized that space operations were a "goal" of the USAF. With the launch of *Sputnik,* these leaders believed that a new dimension of the Cold War had unfolded.

In April 1958, Boushey declared the United States to be "in a race for the control of space."[53] The air force position ran counter to Eisenhower's claim that the United States was not racing the Soviet Union. Hoping to prove that these ideas were not just an air force attempt to squeeze the other services out of the space business, Boushey cited Senator Lyndon B. Johnson as saying that "whoever gains the ultimate supremacy of space, gains control—total control—over the earth for purposes of tyranny or for the service of freedom." Johnson supported the USAF aerospace position.[54]

If the reasoning of White, de Seversky, Boushey, and Johnson was accurate, then the United States was in a life-and-death struggle for the control of space. This sensational idea fueled a call from Gen. Donald Putt, air force deputy chief of staff for development, and Vice Chief of Staff Curtis E. LeMay to weaponize space. Nonetheless, Eisenhower and Secretary of State John Foster Dulles maintained their stance that the use of space was for "peaceful purposes," which included the militarization of space, but not the weaponization of space.[55]

As White, Boushey, Putt, and LeMay campaigned for an increased role for the air force in the use of space, Eisenhower advocated for the creation of a civilian space program and the creation of an agency in the Department of Defense to centralize control of all advanced research projects, especially satellites and ICBMs.[56] Upon the recommendations of the members of the Science Advisory Council and Secretary of Defense Neil McElroy, Eisenhower established the Advanced Research and Projects Agency (ARPA) in November 1958 to work on the military space program of the United States.[57] Also, Eisenhower charged the newly founded NASA with running the civilian space program.[58] It appears that Eisenhower had a different idea from the air force as to the utility of space. Ultimately he devised a plan that charted a middle course leading to the development of a national space policy.

Eisenhower Takes a Different Tack

Attempting to prevent further erosion of America's international image, Eisenhower assigned ARPA the task of streamlining and improving the effectiveness of US military space programs.[59] Despite the air force claim to be the nation's space force, the army and navy continued to work on their own satellites and space systems, which contradicted the idea of centralized control through ARPA.

In November 1957, the army presented a briefing to DoD officials that detailed how its satellite program supported Eisenhower's space agenda. The army's presenter observed that the satellite program could be used to counter increasing international suspicion that the United States was indeed lagging behind the Soviet Union in the space race.[60] The Soviet Union's launch of *Sputnik II* shattered these ideas. The launch of this second Russian satellite definitely shook up the American people, including prominent Eisenhower cabinet members such as John Foster Dulles and his brother Allen.

Throughout the tense months of October and November 1957, the Dulles brothers and Christian Herter continually emphasized the prospective psychological impact of demonstrating America's space capability to the international community.[61] In an effort to deflect the negative impact of the Soviet Union's success with its satellite program, Eisenhower opted to highlight America's overt commitment to NASA and its civilian space program.[62]

The Space Act, favored by Eisenhower and passed by Congress, charged the new space administration with responsibility for the US civilian space program. Eisenhower appointed the president of the Case Institute of Technology, T. Keith Glennan, as NASA's first administrator, believing that Glennan could sway international opinion by using NASA to demonstrate America's superior civilian scientific and technological capabilities.[63] Furthermore, Eisenhower believed that NASA's civilian space programs would reinforce his commitment to using space for peaceful purposes.

In 1958, ARPA and NASA both worked on the military and civilian space programs of the United States. Despite the significant amount of overlap in their work, the two agencies were separate entities. Eisenhower played on this separation to portray the United States as committed to the peaceful research and exploration of space. *The Space Handbook,* written by Robert W. Buchheim of the RAND Corporation, highlights Eisenhower's commitment to exhibit America's interest in the peaceful uses of space. Prepared in December 1958 for the Select Committee on Astronautics and Space Exploration of the House of Representatives, the handbook served as a classified reference work for US politicians.[64] A year later, the RAND Corporation updated the study that was then published by Random House,it providing the American public with a handy reference book on US space programs. The book provided detailed scientific and technological material on US space systems. Furthermore, it reinforced Eisenhower's position on the US commitment to the peaceful uses of space.

Reaffirming its right to use space for national security purposes or military missions, the US government also advocated the peaceful uses of space.

President Dwight D. Eisenhower (center) commissions Dr. T. Keith Glennan (right) as the first administrator of NASA, and Dr. Hugh L. Dryden (left) as the first deputy administrator of NASA, August 19, 1958. NASA, courtesy of nasaimages.org

Supporting this objective, Buchheim, in *The Space Handbook,* wrote: "The statesman, endeavoring to promote world peace, can see both a hope and a threat in astronautics. International cooperation in space enterprises could help to promote trust and understanding. Astronautics can provide physical means to aid international inspection, and thereby, can help in the process toward disarmament and the prevention of surprise attack. Astronautics can also lead to military systems which, once developed and deployed, may make hopes of disarmament, arms control, or inspection more difficult to fulfill."[65]

For the civilian and scientific applications of satellites, the author of *The Space Handbook* mentioned "weather forecasting, aids to navigation and communication, aerial mapping, geological surveys, forest-fire warning, and iceberg patrol." These applications were similar to ones advanced by the Air Force and RAND in 1946–1950.

The classification of "weather, navigation, communication, and mapping" as military applications in 1949–1950 and then civilian ones in 1958–1959 demonstrates that the two US space programs had significant overlap with one another. Despite the overlap, however, Eisenhower maintained their organizational separation.

Beyond civilian applications, Buchheim suggested that satellites offered "novel capabilities" for the military such as "reconnaissance and bombing." In reality, the major military satellite programs begun under Eisenhower encompassed elements useful for both civilian and military applications.[66] The military applications section of *The Space Handbook* supported USAF's belief that control of space was vital in the strategic balance of the Cold War. Even though the air force position on controlling space contradicted Eisenhower's position, the president had a solution. Despite the statements made by high-ranking air force leaders such as White, Boushey, Putt, and LeMay on the need for the United States to control space with the deployment of space-based weapons, Eisenhower remained committed to using space for non-aggressive purposes. Nonetheless, the president recognized that America could not allow the Soviet Union to control space. In a memorandum to James Killian dated February 21, 1958, a White House staffer stated that "the exploitation of unknown area involves two distinct objectives—one, exploration and two, control."[67] In 1958, according to the White House staff member, the "exploitation of outer space fell into civilian-scientific areas rather than military areas," but the military maintained interest in space for "surveillance, communication, and long-range weather forecasting."[68] The military's interest in space as identified by the administration contradicted USAF interest in the weaponization of space. For Eisenhower, the militarization and weaponization of space were two distinct issues. As for the weaponization issue, the February 1958 memorandum to Killian outlined that the issue seemed to be one of "denying access to space."[69] The support of anti-satellite interceptors, to deny the Soviet Union the use of space-based systems, provided Eisenhower with a position on space weapons. For Eisenhower, anti-satellite weapons provided a defensive capability for the United States to maintain space as a weapons-free environment. By claiming that security interests dictated research and development of defensive space systems, Eisenhower and his national security staff moved forward with plans to militarize space, while hedging on the issue of weaponizing space. Eisenhower appeased the air force interest in space weapons by using the perceived threat of the Soviet Union's satellite program to justify America's early work on anti-satellite and ballistic missile defense systems. In the minds of key members of the administration, these two defensive systems would keep space open to all by providing the United States a capability with which to deny the Soviet Union exclusive control of space.

Having temporarily satisfied the air force, Eisenhower solidified his commitment to the peaceful use of space. His next step was to cast the United States as the world leader in the role. To begin, Eisenhower appealed to Senate Majority Leader Lyndon B. Johnson and then to the United Nations.

In late 1958, roughly a year after the launches of *Sputnik* and *Sputnik II,* Eisenhower enlisted the support of Johnson in an effort to build an international consensus on the peaceful use of outer space.[70] Eisenhower extended an offer to Johnson to speak before the United Nations in defense of the US position. Johnson was "happy to accept the invitation in the hope that I can make some contribution toward making outer space an avenue of peace instead of an arena of war."[71] Johnson added that "peace is the first and greatest opportunity space affords."[72] Moreover, in an effort to reinforce America's commitment to working within the international community, Johnson proclaimed that "we need to concern ourselves not with the guarding of what we know, but rather, with exchanging and pooling our knowledge with all the community of free men."[73] Later, as president, Johnson would maintain his advocacy of cooperation in space by signing the Outer Space Treaty in 1967.

Eisenhower's assertion that space should remain an open medium for all to explore and his support of a ban on weapons in space contradicted the belief that the United States had been working since the late 1940s to "arm the heavens."[74] To the contrary, even in the aftermath of *Sputnik* and despite the growing concerns of the air force and aerospace advocates about the need to weaponize and secure space, Eisenhower reaffirmed America's open commitment to the peaceful use of space. The formation of NASA and the initiation of an international dialogue on the banning of weapons from space supported the conclusion that the United States did not seek to arm the heavens.[75]

This was not a foreign concept during the Cold War. In the aftermath of World War II, the US representative to the United Nations, Bernard Baruch, proposed the creation of an international commission "for the control of atomic energy to ensure its use for peaceful purposes only."[76] Although the Soviet Union rejected the Baruch Plan because they thought it was a vehicle for American hegemony, the Eisenhower administration resurrected the essence of Baruch's proposal in the months after the launch of *Sputnik.* Secretary of State John Foster Dulles announced at a news conference on January 16, 1958 that the United States supported the creation of an international commission within the United Nations to ensure that the exploration of space be conducted "in the interests of science and humanity, and not in the interest of war."[77] Dulles referred to the Baruch Plan as a "missed opportunity," and challenged the Soviet Union to join the United States to work toward the preservation of space as a weapons-free frontier. The Dulles speech built upon an early UN disarmament proposal that addressed the issue of the use of space for peaceful purposes; the proposal, led by the United States, France, Canada, and Great Britain, strove to build international support for this idea.[78] Eisenhower used the growing international interest in preserving peace in frontier areas such as Antarctica and space to cast the United States as the leader in the international community.[79] Even his "Open Skies proposal" fit into this paradigm of gaining international prestige for the United States by advocating an international agreement designed to reduce tension within the global security environment.

Committed to the belief that space needed to remain a weapons-free frontier, Eisenhower nonetheless supported the militarization of space. Although Eisenhower's position at first seems contradictory, America's military space programs focused on the use of passive and non-aggressive satellites for communication, surveillance, and weather forecasting missions. The non-aggressive militarization of space allowed Eisenhower, and future Presidents Kennedy and Johnson, to argue that the US military satellite programs maintained stability and security in a tense period of Soviet militarism and communist expansion. Eisenhower and future US presidents came to argue that possessing non-aggressive passive intelligence gathering satellites allowed the United States and the free world to peer behind the Iron Curtain with an eye toward maintaining international stability. For Eisenhower, satellites provided an advanced method of verifying the military disposition of the Soviet Union and allowed the United States to protect itself, while also preserving the new frontier for the benefit of the international community.

The advocacy of space-based weapon systems withered as support for research into ASAT and BMD systems blossomed. For Eisenhower, deploying ASATs and BMDs contradicted his fundamental commitment to the principle of space as a peaceful frontier. By choosing not to deploy space weapons, Eisenhower reaffirmed his position that a single nation should not control space.

The Militarization of Space

To Eisenhower and his subordinates John Foster. Dulles, Thomas White, and Homer Boushey, *Sputnik* mandated an American response, though not necessarily a symmetrical one. Not wishing to outright race the Soviet Union in civilian "space stunts," Eisenhower focused on developing military satellite systems designed to improve America's intelligence gathering and early-warning capabilities.

Dedicated space-based early-warning systems emerged from ARPA as well as USAF. After assessing the service's military space programs in January 1958, the staff of ARPA presented a list of space systems and projects "essential to the maintenance of our national position and prestige."[80] The five systems and their numerical designations listed by ARPA in the beginning of 1958 were "609 Ballistic test and related systems, 447 Manned Hypersonic Research Systems, 463 Dyna-Soar, WS-117L Satellite System, and 499 Lunar Base."[81] Once ARPA identified the major systems and subsystems, they distributed programs and development of the components among the services, despite the air force claim that it was the nation's space force. In June 1958, the three services had transferred their space work to ARPA, even though the air force went ahead and established its own Directorate of Advanced Technology.[82] In summer 1958, ARPA apportioned the military space systems research to the three services. According to air force historian Lee Bowen, the USAF received the "weapons system to control hostile satellites, nuclear bomb propelled space craft, WS-117L, lunar probes, re-entry studies, and project SCORE," while ARPA gave the US Army the research programs on the "1,500,000-pound booster, the meteorological satellite, and inflatable sphere."[83] Meanwhile, the US Navy remained committed to research on the use of satellite navigation and communication, but used laboratories such as the Applied Physics Laboratory (APL) at the Johns Hopkins University for research and development of satellite systems and advanced weapons. America's military space programs were well underway. The satellite and space-based systems developed in this period formed the foundation of the US military space program.

WS-117L became the first large-scale satellite program designed specifically for the military's use. Although USAF instituted the program in 1956, the service suffered through the transfer of the program to ARPA at Eisenhower's order. After analyzing the progress of the WS-117L program in comparison to research and development conducted by the US Navy on its satellites, ARPA found no duplication between the programs and returned the WS-117L to the air force.[84]

The WS-117L program encompassed more than just a reconnaissance satellite. Committed to avoiding a "nuclear Pearl Harbor," Eisenhower ordered the Technology Capabilities Panel in 1955 to study how to use advanced technologies to reduce America's risk from a Soviet surprise attack. The emergence of Soviet ICBMs in the mid-1950s further fueled the president's concern about preventing a Soviet attack. The members of the TCP recognized satellites as a promising technology, useful in the design of a missile early-warning system.[85] With the launch of *Sputnik* atop a Soviet R-7 ICBM, Eisenhower's concerns over the emerging Soviet ICBM threat compounded. America needed an advance early-warning system. The Missile Defense Alarm System (MIDAS) evolved out of this need. Classified as Subsystems G (missile detection component) of the WS-117L program, MIDAS was to be a space-based detection network for the United States.[86]

Although work on the Distant Early Warning (DEW) system and the more capable Ballistic Missile Early Warning System (BMEWS) continued, these ground-based radars provided at best a fifteen-minute warning of an incoming Soviet ICBM attack.[87] Air Force leaders, especially within SAC, championed the idea of using a space-based missile detection network because it could provide US commanders with an additional fifteen minutes of strategic warning.[88] In simple terms, a space-based missile detection network could provide the United States with as much as thirty minutes of advance warning of a missile launched from the Soviet Union. Based on military response scenarios envisioned by SAC during the Cold War, the additional fifteen minutes allowed a greater degree of flexibility in response scenarios for US strategic force commanders.[89] SAC lobbied the administration on the benefits of using a space-based early-warning system in conjunction with the ground-based detection systems.

The air force lobbying effort paid off in September 1959 when Secretary of Defense Neil McElroy reassigned MIDAS, the missile defense alarm system, from ARPA to the air force. Within a year, USAF, with assistance from Lockheed program managers, had two MIDAS test satellites ready for evaluation.

In addition to the development of MIDAS, the air force worked on the observation component of WS-117L that evolved into the Satellite and Missile Observation System (SAMOS) program. The air force envisioned SAMOS as a photographic and electromagnetic data collection vehicle, but the SAMOS satellite eventually lost out to the CIA's Corona photoreconnaissance satellite program. Since 1951, when C. P. Cabell first outlined the requirements for a photoreconnaissance satellite to Bernard Schriever, USAF had been working on solving technical issues associated with developing cameras, film, and hardware for use in space.[90] Corona represented the triumph of these early efforts. Because the original WS-117L program included a photoreconnaissance system, the CIA became interested in the program in late 1957.[91] Compared to the advanced designs of SAMOS, the photoreconnaissance element of WS-117L offered much greater capabilities. In spring 1958, the photoreconnaissance system of WS-117L was re-designated Project Corona[92] The CIA and the air force worked together on the new project to develop a reconnaissance platform beyond the secret U-2 program.

The importance of space-based reconnaissance increased as Eisenhower learned of the vulnerability of U-2 spy planes flying over the Soviet Union.[93] On May 1, 1960, Eisenhower's concerns about the Soviet Union's ability to intercept the U-2 became justified with the downing of Francis Gary Powers over the Soviet Union in a CIA U-2.[94] Space-based systems won additional support from Eisenhower in the ensuing muddle following the downing of Powers. In August 1960, just three months after the May Day incident, the first successful Corona mission lifted off from American soil. On August 19, 1960, Eisenhower and Allen Dulles inspected the first photographs of the Soviet Union taken by a satellite. The new space-based photoreconnaissance system worked well and offered several advantages over the U-2. The fiasco of the May Day incident and the success of Corona in August 1960 enhanced Eisenhower's commitment to the use of non-aggressive military missions in space. After years in research, development, and engineering, the Corona satellite finally provided data about the Soviet Union's strategic and military capabilities. Space had become militarized.

In addition to the MIDAS, SAMOS, and Corona satellite systems, Eisenhower welcomed the development of weather, communication, and navigation satellite systems such as TIROS and Transit that provided potential benefits to the international civilian community as well as to the US military. TIROS (Television Infrared Observation Satellite), a NASA program by 1960, was the culmination of the feasibility study done by Greenfield and Kellogg in 1951. Despite control by NASA, TIROS was developed with significant assistance from the US States Army Materiel Command.[95] NASA claimed that the TIROS satellite was the "world's first meteorological satellite" designed to provide "cloud pictures."[96] The use of satellites to acquire weather data evolved from the initial feasibility study done by RAND in 1946. However, despite the initial classification of weather satellites as military systems in the late 1940s and early 1950s, Eisenhower recognized that weather satellites could be beneficial to the international civilian community as well as the US military services. While the military services and NASA maintained independent weather satellite programs, Eisenhower used

Engineers work on TIROS I satellite, April 1, 1960. NASA, courtesy of nasaimages.org

NASA's TIROS to legitimize his claim that the United States supported only the peaceful use of space.[97]

NASA allowed the public to see the details of TIROS through the publication of brochures, as well as allowing reporters access to the weather satellite system. NASA encouraged publicity to reinforce its open nature. The openness of the TIROS program stood in stark contrast to the secretive Corona program.[98] Weather satellites such as TIROS served Eisenhower both as a vehicle for the collection of vital weather data for the planning of strategic operations and as a way of giving the press something to focus on instead of snooping around for information about Corona.

In addition to TIROS, the US Navy provided Eisenhower with the Transit satellite system in 1959, calling it the "world's first navigational satellite

First photograph of Earth from TIROS I satellite, April 1, 1960. NASA, courtesy of nasaimages.org

system." Transit[99] originated with the launch of *Sputnik* and the work of scientists George Weiffenbach and William Guier. Weiffenbach and Guier worked for the APL at Johns Hopkins University. Intrigued by *Sputnik,* the two scientists worked to develop a more precise method of locating its position.[100] They noted that they could determine the position of *Sputnik* by "analyzing the Doppler shift in its radio signals." Frank McClure, chairman of the Research Center of the APL, observed: "If the satellite's position was known and predictable, the Doppler shift could be used to locate a receiver on earth." The US Navy, which had a close working relationship with APL, welcomed the idea. In 1958, the Transit program began.

The US Navy's initial interest in the satellite navigation project emerged as a way for Polaris submarine crews "accurately to determine their position while on extended patrols."[101] The use of navigation satellites provided the

Ground terminal for Relay communication satellite, December 13, 1962. NASA, courtesy of nasaimages.org

US Navy and even USAF with another non-aggressive, yet military use of space. Like TIROS, Transit began as a military program and evolved into a civilian system. Three years after the initial operational capability (IOC) date of Transit, the US government opened the system to limited civilian use. Despite the valuable military potential of Transit, in 1967 President Johnson opened the satellite navigation network for "world-wide commercial use." Again, the United States developed dual-use satellites that provided valuable military and civilian benefits. TIROS and Transit exemplified Eisenhower's commitment to the military and civilian uses of space.

In addition to the use of space for weather and navigational purposes, the military services developed passive and active communication satellites. Programs such as Echo, SCORE, and Courier provided America's military with their initial space-based communications systems. The TCP recommended

Echo satellite inflation test at NASA's Langley Research Center, August 12, 1960. NASA, courtesy of nasaimages.org

in 1955 the use of communication satellites for secure international military radio traffic. The civilian communications community also recognized the vast potential for the use of space-based satellites for radio and television broadcasts.[102]

The first communication satellites, such as Echo, were nothing more than large self-inflating Mylar spheres that the military and NASA launched into low earth orbit. After the sphere had attained it proper size, it acted as a relay point to "bounce" radio signals to distant ground-based transceivers.[103] After experimenting and using "passive systems," the military began work on advanced repeater and transceiver satellites.

In December 1958, Eisenhower publicly praised the success of the first repeater satellite, SCORE. Project SCORE began in response to the Soviet Union's successful launches of *Sputnik* and *Sputnik II.* Eisenhower authorized the use of an Atlas booster to place a rudimentary repeater satellite, one with a prerecorded Christmas message for the world, in low earth orbit. Project SCORE not only demonstrated America's space capabilities but also, as an active transmitter, was a step in the evolution of communication satellites beyond the passive models like Echo. Furthermore, Eisenhower's authorization of a military booster for SCORE contradicted his early position concerning Project Vanguard that civilian programs had to use civilian boosters. In the end, Project SCORE demonstrates that civilian and military space programs closely overlapped despite Eisenhower's public insistence that they were two separate programs.

After the success of Project SCORE, the administration approved Project Courier, an advanced active communications satellite, in July 1959.[104] Whereas SCORE was a repeater satellite model in which a prerecorded message was loaded into the satellite's software and then transmitted to ground-based receiver stations, Courier promised a more sophisticated technology for transmitting messages.

Along with the military applications, the civilian market provided the impetus for US businesses to pursue and design advanced communication and television satellites for commercial use.[105] These commercial ventures supported Eisenhower's commitment to the open use of space by the international community. Eisenhower used both military and civilian communication, navigation, and weather satellite systems to combat the success of the Soviet Union. He believed that these satellite systems demonstrated America's scientific and technological expertise to the international community. The USAF still openly advocated the weaponization of space but Eisenhower remained committed to his policy position of keeping space free of weapons. He rejected the use of moon bases for missiles, active space shields, and orbiting bombardment satellites, in favor of research into ASAT and BMD systems.

The thought of intercepting hostile satellites emerged in the United States within days after the launch of *Sputnik*.[106] Air Force officials such as White, Boushey, and Schriever all recognized that control of space had strategic significance. As engineers and scientists completed work on the use of satellites for early warning and reconnaissance, the need for a satellite that could "kill" an enemy's space-based system emerged in the wake of *Sputnik*. The initial identification by the air force of the need for space-based ASAT systems clashed with Eisenhower's public commitment to use space for pacific purposes.

Initially USAF envisioned the use of satellites much like MIDAS or SAMOS to intercept Soviet satellites by using either a nuclear warhead or kinetic energy. Despite USAF insistence on the need for such a program, Eisenhower shied away from space-based ASAT systems and initially confined ASAT research to "studies only." Moreover, he wished to keep the public unaware of the program. As the weaponization debate raged within the military services between 1958 and 1960 over the need for the United States to develop a satellite interceptor (SAINT), Eisenhower remained committed to keeping space free of weapons.[107]

By the beginning of 1960, the debate over ASAT had reached fever pitch in the administration. In a meeting of the NSC on February 5, 1960, the assistant secretary of the air force, Dr. Joseph Charyk, took a new tack in seeking to win NSC and presidential approval of SAINT.[108] Charyk emphasized the inspection element of the SAINT system as a way for the program to fit within the president's policy vision. The president's science advisor, George Kistiakowsky, in the same NSC meeting on February 5, 1960, presented arguments against the demonstration and use of ASAT technology by the United States. Kistiakowsky recounted in his diary that "the President rather unemphatically agreed with my position, but it didn't sound as if it was a directive."[109] Even after the debate on ASAT systems during the NSC meeting, Eisenhower remained devoted to keeping weapons out of space. Despite the obstacles raised by the president and Kistiakowsy, by the end of Eisenhower's tenure there were increased concerns over the Soviet Union's ability to orbit bombardment satellites, later prompting John F. Kennedy to support the development of US ground-based ASAT systems.

Having confined ASAT systems to "research studies," Eisenhower authorized the investigation of technologies for the development of a ballistic missile defense system.[110] However, only Operation Argus and Project Defender directly entailed the use of space for the interception of hostile ballistic missiles.

Operation Argus, operated by ARPA, evolved from a theory postulated by Nicholas C. Christofolis of the Lawrence Livermore National Laboratory. Christofolis hypothesized that high-altitude nuclear detonations released an "increased amount of electrons" that generated a large electromagnetic pulse (EMP).[111] Christofolis theorized that EMP could provide a space shield against incoming ballistic missile attacks.[112] Between August 27 and September 6, 1958, ARPA with assistance from all three military services conducted high-altitude atomic bomb tests to confirm Christofolis's hypothesis, which turned out to be correct. However, after the course of tests, ARPA and the services noted that the temporary shield "dissipated rapidly" and thus would be of little value in the interception of incoming ballistic missiles.[113]

Despite the limits of the success of Operation Argus, ARPA continued its investigation into ballistic missile defenses. Project Defender was another ARPA space-based BMD program. Project Defender encompassed numerous

BMD programs, one of which was Project BAMBI (ballistic missile boost intercept). BAMBI was a program designed to intercept enemy ICBMs in their boost phase, generally recognized as the easiest phase to intercept a missile. Critical to BAMBI's mission was the use of satellite tracking to identify a hostile missile and then relay interception data to an intercepting missile. This technology paralleled the work done in the MIDAS and SAMOS programs. Unlike the use of ASAT weapons, which occasioned much debate, the use of space for missile defense and vital intelligence gathering did not openly contradict Eisenhower's pledge to keep space peaceful.

The Space Patrol Active Defense (SPAD) program offered a follow-on technology to the original BAMBI system. ARPA scientists believed that space-based interceptors could be stationed in orbit and used to attack incoming Soviet ICBMs with large "wire mesh" obstacles. While technically feasible, SPAD violated the president's ban on space weapons; therefore, it remained a research program. Despite the ideas and designs developed and envisioned within Project Defender, space remained weapons-free during Eisenhower's tenure.

The promise of systems such as SAINT, BAMBI, and SPAD failed to resonate with Eisenhower. He remained committed to the militarization of space without advocating its weaponization. By the end of his second term as president, the United States used satellites to gather military intelligence, collect weather data, and transmit communications, but had no weapons stationed in space. Eisenhower had remained true to his pledge to preserve space as a weapons free-frontier, while aggressively pursuing the militarization of space.

Conclusion

Despite the fact that historians largely credit Eisenhower with ushering the United States into the space age, he built on the foundation of research established in the years 1946 to 1952. In contrast to Truman, Eisenhower seized the initiative to take advantage of space for military purposes, worked to keep space free of weapons, and championed the establishing of NASA. Using the ideas, feasibility studies, and research done during Truman's administration, USAF and the RAND Corporation served as leaders in the movement toward the militarization of space. Despite the continual refrain from RAND employees and USAF space advocates that the United States would benefit greatly from the international prestige of being the first nation to launch a man-made satellite, Eisenhower shied away from "space stunts" in order to concentrate US resources on the development of intelligence-gathering satellites. *Sputnik* and the evidence that the Soviet Union had operational ICBMs in mid-1957 pushed Eisenhower to support programs designed to meet the nation's need for space-based intelligence and reconnaissance vehicles. Unlike the use of aircraft such as the U-2, satellites programs such as Corona, MIDAS, and SAMOS provided an alternative to invading the Soviet Union's air space with manned missions, and supported Eisenhower's failed proposal of "Open Skies." Under Eisenhower, satellites were pressed into reconnaissance service, as well as early-warning and sentry duties deemed vitally important in the Cold War. These efforts supported Eisenhower's New Look policy because they provided an advanced technological solution that de-emphasized the production of costly manned equipment and showcased America's technological capabilities. Eisenhower's New Look policy not only reshaped the structure of the US military, but also fundamentally influenced his support of military and civilian space systems. His support of the non-aggressive military use of satellites formally advanced the belief that space should be preserved for peaceful exploration, which the newly created NASA pursued under its charter.

Contrary to the goals and missions of NASA, airpower advocates clamored over the Soviet Union's moves to control space, and USAF pressed for the weaponization of space. Although Eisenhower dismissed concepts such as missile bases on the moon and armed satellites, he entertained research on the use of ASAT and BMD systems. While the NSC staff debated anti-satellite systems such as SAINT, Eisenhower steadily stressed that he wished to avoid space warfare. At Eisenhower's instruction, ASAT and BMD weapons remained research projects, much as satellites had under Truman.

The decision to keep space peaceful provided Presidents Kennedy and Johnson with a foundation on which to continue work toward the completion of ground-based ASAT and BMD systems, while also honoring Eisenhower's pledge. As Eisenhower had benefited from the satellite research started under Truman, Kennedy and Johnson built upon the ideas and programs supported by Eisenhower. The establishment of NASA and the development of dual-use satellites such as Transit and TIROS offer evidence that Eisenhower was not interested in the conquest of space. While NASA had ties with the military and often worked closely with the services, Eisenhower believed the space administration exemplified his commitment to the peaceful exploration of space. The Transit and TIROS satellite programs ultimately provided navigation and weather technologies beneficial to the international civilian community.

Although Eisenhower supervised the development of two distinct and at times seemingly contradictory space programs, he never wavered from his commitment to the peaceful use of space. While the Eisenhower era is often characterized by policies such as massive retaliation and New Look, the period 1953 to 1961 needs to be remembered as the era in which the United States steadfastly committed itself to the non-aggressive militarization of space, NASA's civilian space program, and the goal of banning weapons from space. The impact of Eisenhower's actions resulted in a solid commitment to a national space policy that emphasized the use of the military and civilian space programs working in concert to reestablish the technological superiority of the United States.

Chapter 3: Kennedy, Disarmament, and FOBS

While maintaining the key military space programs begun during Eisenhower's presidency, John F. Kennedy strove to demonstrate the US commitment to peaceful exploration and non-aggressive military uses of space and built on Eisenhower's pledge to keep space free of weapons. To demonstrate, Kennedy capitalized on the Soviet Union's expressed interest in orbital bombardment satellites by fundamentally shifting the public's attention to NASA's civilian programs, while also working to ensure that space remained free of nuclear weapons. With his 1961 pledge to land a man on the moon by the end of the decade, Kennedy singlehandedly captured the imagination of the international community. Although Kennedy moved away from Eisenhower's notion that the United States should not engage the Soviet Union in a competition of "space stunts," he believed that the United States could gain a vital psychological advantage in the international strategic balance by emphasizing the US civilian space programs.[1] Kennedy envisioned using NASA as a showcase to demonstrate the peaceful scientific, technological, and engineering capabilities of the United States, while he depicted the Soviet Union as an aggressive enemy bent on control of space.

Kennedy used two approaches to further the US objective of preserving space as a weapons-free frontier. First, he established a relationship with Soviet Premier Nikita Khrushchev so that the two leaders could work toward an international agreement. Recognizing that a space disarmament agreement would require a long, tedious process, Kennedy nonetheless believed one could be reached and favored using the United Nations to help accomplish this. Second, Khrushchev's constant rhetoric about the Soviet Union's ability to orbit weapons in space concerned Kennedy and his national security staff and led to the president's support of ground-based anti-satellite systems. The thought of the Soviet Union controlling space led Kennedy to authorize the air force and army to develop ASAT systems capable of "neutralizing" hostile enemy satellites or orbiting weapon stations. The concern over the Soviet Union's apparent bid to conquer space, not exclusive to the Kennedy administration, had been a constant issue since the launch of *Sputnik* in

1957. By supporting the development of an operational ground-based ASAT system, Kennedy strove to provide the United States with the capability to intercept the Soviet Union's orbiting weapons system and demonstrate that the United States was committed to keeping space free of weapons. The idea of ASAT systems originated under Eisenhower, but Kennedy expanded the programs beyond the R&D phase.

Kennedy believed that ASAT systems did not violate America's pledge to maintain space as a weapons-free environment. He argued that they were defensive weapons developed to keep space free for all nations and that the United States had the right to preserve its national security.[2] To counter the Soviet Union's rhetoric about the need to ban all military uses of space, Kennedy defended the US military's use of space for reconnaissance, weather data collection, and communication as justified for national security and stability in the international security environment.[3]

The significance of Kennedy in the history of the development of a national space policy is that he continued to advance the military space programs begun under Eisenhower while also seeking to reach an arms control agreement with the Soviet Union. To reinforce his position, Kennedy highlighted and expanded America's civilian space program as a gesture to demonstrate America's commitment to the peaceful exploration of space.

Senator Kennedy and Space

John F. Kennedy was the first US president to have dealt with space issues prior to entering office. Although not directly involved in the post-*Sputnik* hearings as a US senator from Massachusetts in the 1950s, Kennedy answered a flurry of letters from constituents in the aftermath of the Soviet Union's launch of *Sputnik*. A major concern of Kennedy's constituents was the position of the United States in relation to the Soviet Union. Essentially Kennedy answered the same questions that the public put to Eisenhower.

The Soviet Union's launch of *Sputnik* had shocked the American public in 1957. After a decade or more of believing that the United States was the world's leader in science, technology, and engineering, the US public had a difficult time accepting that a "backward" communist country had beaten the United States into space.[4] The public wanted answers. While Eisenhower de-emphasized the significance of *Sputnik,* the election of 1960 focused the public's interest on the "space race." As he emerged as the Democratic candidate for the presidency in 1960, John F. Kennedy increasingly fielded questions about this topic.

In January 1960, a private citizen—Theodore Brassord Jr. of Taunton, Massachusetts—wrote to Senator Kennedy about America's space program. Brassord said that the real issue at stake in the space race was the "control of space," and he advised Kennedy to change the US system for funding space research, calling it "weird."[5] Kennedy responded by stating that the US space program was anywhere from "two to six years behind the Russian program." He asserted that the lag resulted from "organizational issues, leadership, and planning." No doubt, Kennedy was blaming Eisenhower and his administration for falling behind the Soviet Union in the space race, but Kennedy remained silent as to Brassord's thoughts on "control of space." Kennedy's bland, generic reply to Brassord stands in stark contrast to his response to a Princeton University freshman a month later.

In February 1960, Senator Kennedy received a letter from William Everdall—a freshman at Princeton University and self-identified Republican—who wanted Kennedy to advocate "more money for the space program," as well as to streamline the space bureaucracy.[6] Everdall believed the US space program had "unlimited glamour and incredible implications." Kennedy replied to Everdall's letter in greater detail than he had to Brassord's. Outlining his "views on U.S. space policy," Kennedy stated:

> Whatever the scale and pace of the American space effort, it should and is a scientific program. In this interval when we lack adequate propulsion units, we should not attempt to cover this weakness with stunts. And when this weakness is overcome, our ventures should remain seriously scientific in their purposes.
>
> Since the exploration of space is, scientifically, a relatively new venture, it is rational to expect pay-offs we cannot calculate, as in the early stage of any major scientific breakthrough. This has two consequences. First, the basic scientific component of our program should be financed and encouraged to the hilt. It is out of the work in basic research that possibilities of leap-frogging the Russians are likely to emerge. And without leap-frogging I fear we shall be getting their exhaust in our face for quite a long time. Second, projects for exploration should at this early stage be viewed with a bias toward hope rather than skepticism. We can count on good payoff from a high proportion of our probes, at this stage. Thus, on a scientific basis alone, the program should be generously financed.
>
> With respect to the competitive and psychological aspects of the space program, it is evident that we have suffered damage to American prestige and will continue to suffer damage for some time.[7] But, our recent loss of international prestige results from an accumulation of real or believed deficiencies in the American performance on the world scene: military, diplomatic, and economic.

> It is not simply a consequence of our lag in the exploration of space vis-à-vis the Soviet Union. The space lag has in fact, had a disproportionate impact because it is one of a group of lags and gaps.[8]

In his long response to Everdall, Kennedy stressed the importance of maintaining the civilian emphasis of America's space program at a time when President Eisenhower believed that the military elements were of equal or greater value. Despite that difference of opinion, Kennedy initially agreed with Eisenhower that the United States should not engage in "space stunts" to appease the public's anxiety, although this was a position he eventually reversed as president.[9]

Kennedy's reply is important because it provides insight into his thinking on the US space program, and because it is a veiled critique of the Eisenhower administration's space policy. As the 1960 presidential election loomed, Kennedy intensified his criticism of the administration's management of the US space program since he believed that the United States had fallen behind in the international community's perception of the space race.

Supporters encouraged Kennedy to increase his attacks. On September 7, 1960, the director of the Harvard College Observatory, Donald H. Menzel, sent Kennedy a sixteen-page memorandum highlighting Eisenhower's faults in handling the US space program. Menzel claimed that the Eisenhower administration focused too much on "technology and engineering" and seemingly ignored the need for "basic science."[10] Menzel maintained that "slashed budgets and lack of science education in America" resulted in the second-place status of the United States in the space race. Menzel squarely blamed the president although Eisenhower and his staff had themselves identified similar issues in the aftermath of *Sputnik.* In the closing months of the 1960 campaign season, Menzel advised Kennedy that the Democrats needed to "impress" US scientists by reinvigorating research funding for basic science. Space and science, from Menzel's point of view, had become major issues in the 1960 election.

Contrary to Menzel's accusations, Eisenhower and NASA administrators had not neglected basic science. Eisenhower worked to fully fund both the military and civilian space programs after 1957. On the civilian side, NASA worked diligently on manned space flight, as well as weather and communication satellites, in an effort to demonstrate America's commitment to science and technology. On the military space programs, Kennedy, while senator, remained strangely silent while Eisenhower authorized missions with Corona and development of the MIDAS, SAINT, and Vela Hotel satellites. Instead,

Kennedy focused on Eisenhower's perceived lack of leadership and vision for the US space programs, believing that the United States had suffered an international setback because of the Soviet Union's space successes.[11] In an interview for the magazine *Ground Support Equipment,* Kennedy reinforced his position, stating: "We must make every effort to halt this Soviet propaganda drive; we must reassure the non-communist nations of the world that America is strong enough to meet its commitments to world peace. We can do this only through our achievements."[12] Suggesting the seriousness of the space race, Kennedy maintained, "[W]e are in a strategic space race with the Russians and we have been losing." He stressed that the United States could not afford to be "second in this race" because "space is our great new frontier." Rather than advocating that the United States unilaterally control space, he recommended working with the United Nations to assure the "freedom of space" for all nations.[13]

Kennedy believed his position presented a strong international image for the United States by championing the civilian exploration and peaceful uses of space, rather than arming the heavens. In Kennedy's thinking, space and satellite technology were vehicles to rebuild America's lost prestige. To achieve his objective, Kennedy focused the public's attention on the civilian space program. He believed NASA offered the best means to demonstrate the scientific and technological capabilities of the United States, as well as to reinforce his stated intention of working to keep space free of weapons.

Kennedy presented his ideas about the US space program before Congress soon after becoming president. Addressing Congress on May 25, 1961, with a special message regarding national space policy, Kennedy stressed the bipolar Cold War realities of the period and stated that space exploration "was not merely a race" with the Soviet Union but an endeavor through which the United States could preserve the freedom of space. He also outlined his goals for the space program. First, he recommended that the United States commit to "landing a man on the Moon and returning him safely by the end of the decade."[14] Second, Kennedy asked for an additional $23 million to fund the Rover nuclear rocket. Third, Kennedy asked Congress for an additional $50 million to accelerate the communication satellite programs designed to enable global communication. Finally, Kennedy asked Congress to authorize an additional $75 million for weather satellites.

With the exception of the moon mission, the majority of the space programs advocated by Kennedy had already been under development during the Eisenhower administration, and Kennedy merely wanted to continue funding the projects already underway. By bolstering US investment in the

Pres. John F. Kennedy's speech before a joint session of Congress, May 25, 1961. NASA, courtesy of nasaimages.org

space program, Kennedy wanted to show the international community that the United States would honor its pledge to maintain the non-aggressive uses of space and lead the international community in the peaceful exploration of space. He also wanted to work toward an international arms control treaty for space.

The following year in a speech at Rice University in Houston, Kennedy continued his effort to get the public to perceive US scientific and technological might. Before a large and distinguished audience that included Vice President Lyndon B. Johnson, NASA Director James Webb, and prominent aerospace engineer Larry Bell, President Kennedy said that the United States would become the "world's leading space-faring nation" by "choosing to go to the moon."[15] Kennedy justified his decision to go to the moon because he believed that "there is new knowledge to be gained, and new rights to be won, and they must be won and used for the progress of all people." In advancing this challenge, Kennedy proclaimed: "There is no strife, no prejudice, no national conflict in outer space as of yet. Its hazards are hostile to us all. Its conquest deserves the best of mankind, and its opportunity for peaceful

President Kennedy speaks at Rice University, September 12, 1962. NASA, courtesy of nasaimages.org

cooperation may never come again."[16] Kennedy firmly believed that space provided a medium for the international community to work beyond the constraints of the bipolar tension of the Cold War.

In his first year in office, Kennedy strove to improve America's international prestige by investing heavily in the manned lunar mission. Despite Kennedy's rhetoric of peaceful and cooperative space exploration, the Soviet Union posed a significant threat to his vision. To insure that the United States remained at the forefront of promoting space for peaceful purposes, in May 1962 Kennedy established the National Security Action Memorandum (NSAM) 156 Committee, an ad hoc group of senior advisors to provide guidance and oversight for the administration's development of space policy.[17] The NSAM 156 Committee focused on making sure that the United States retained "freedom of action" over the use of space for peaceful purposes, including the use the reconnaissance satellites, as the administration worked with the United Nations and the Soviet Union to ban the weaponization of space.

The Military and Space

Based on their strategic concerns about the Soviet Union's space program and strategic rocket force, Kennedy and his national security staff recognized the continued importance of the US military space program, but publically highlighted NASA over the military's space programs. Since the launch of *Sputnik,* US military officers had worried about the Soviet Union's attempt to control space. During Kennedy's presidency, the chairman of the Joint Chiefs of Staff, General Lyman L. Lemnitzer, spoke at the Key Issues Forum in Philadelphia on October 18, 1961. He stated there that the United States "faced a global struggle" in which the "Communists have announced their objective of world domination."[18] Highlighting the encroachment of the Soviet Union into Latin America, Cuba, and Southeast Asia, Lemnitzer acknowledged that the Soviet Union's success in space hinted at their designs to control space. Lemnitzer said that he believed that the Soviet Union's "*Sputnik* and manned orbital flight through space" had direct "military implications." He contended that the Soviet Union's "advances in science brought decisive strategic advantages" and asserted that the Soviet Union's success in space also provided a significant strategic advantage over the United States. Lemnitzer advocated developing "an operational defense against missiles" as a way to offset the lopsided strategic balance. But he was wrong. Concern that the Soviet Union's scientific and technological advances led directly to "strategic advantages" was not new; in fact it had been a continual concern within the US national security apparatus since the Air Policy Commission under President Truman. Moreover, the Soviet Union did not possess an advantage in the strategic nuclear balance. With his comments, Lemnitzer merely restated an already accepted position. Military leaders remained deeply concerned about Kennedy's support of NASA because of their perceived ideas about the Soviet Union's conquest of space. However, Kennedy did not neglect the military's space program, but maintained it while striving to improve the waning international prestige of the United States by supporting NASA.[19]

In contrast to Lemnitzer's advocacy for a BMD system, the air force did not openly advocate the stationing of weapons in space.[20] Instead they remained primarily concerned about the need to improve the international image of the United States. According to Air Force historian Carl Berger, the paramount USAF concerns were the Soviet Union's lead in space flights, weaknesses in the US space program, and the international prestige of the United States. He contended that the air force position matched Kennedy's. The air force used the Soviet Union's successful flights of Yuri Gagarin in April 1961 and German S. Titov in August 1961 as catalysts to lobby for a larger air force "role in the U.S. civilian space program."[21] This was not a mere repeat of Lemnitzer's argument, but a concerted effort by the air force to offer the Kennedy administration a re-conceptualized national space policy led by USAF, not NASA.

Even before Kennedy's inauguration, the air force moved to improve its status in the US space programs by advocating that the United States regain its reputation for scientific and technological leadership within the international community. Believing that Eisenhower had stifled the USAF role in the overall space programs of the United States, air force leaders renewed their hopes for a reversal of policy with Kennedy.[22] The deputy director for war plans, Brig. Gen. J. D. Page, outlined the air force position on space for the new administration. Page maintained that there was no need for a separate space command as proposed by the army and navy. Rather, Page argued that the air force's concept of aerospace remained a useful way of thinking about space operations. Furthermore, he held that US space requirements ought to be considered in the context of how they assisted the services in their defined missions.

Page suggested that the National Aeronautics and Space Act of 1958 be amended to recognize the military's role in space, rather than just focusing on "space for peace."[23] The press characterized this concerted effort by the air force as a "political offensive" designed to reorient the US space programs. Defending the USAF position, the USAF chief of staff, Gen. Thomas White, reassured Congress and the public in December 1960 that the air force did not plan to "undercut" NASA's role in the US space programs.

To clarify the air force position on space, Maj. Gen. Donald A. Burchinal, the director of plans, sought in December 1960 to create an "official view of the future of the Air Force in aerospace [operations]." In January 1961, a draft of the document went to the air force vice chief of staff, Gen. Curtis E. LeMay, who forwarded the draft to the Office of the Secretary of the Air Force (OSAF) for review and approval. The air force officially accepted the new document on June 19, 1961, as Air Force Objective Series (AFOS) 2/2, *Long-Range Concepts as to the Nature of Future War: USAF Views on Military Activities in Space.*[24]

AFOS 2/2 was not a revolutionary document. Rather the document reinforced the air force commitment to the concept of aerospace. AFOS 2/2 also asserted that "national power in peace and war would be greatly influenced and might be ultimately determined by space capabilities." Both of these ideas originated in the late 1940s, and by 1961 had become entrenched in the institutional thinking of the air force. In recognizing the need for continual R&D, the air force outlined its priorities in AFOS 2/2. Specifically, the air force saw "manned exploration of space, maneuverability, low-cost boosters, and aerospace medicine and life sciences" as the main efforts of the national space program. NASA had the same research priorities.[25]

Expanding its vision for the future space policies of the United States, the air force ("Ad Hoc Task Group of the Air Staff") authored *The Space Plan* as a follow-on document to AFOS 2/2.[26] *The Space Plan* became the USAF map for future US space programs, while also positioning the air force as the nation's leader in space. Within the context of *The Space Plan,* the air force openly contended that the separation of the civilian and military space programs was "absurd."

In the introduction of *The Space Plan,* the authors stated:

> The military superiority of the United States and the Free World has been achieved and maintained by virtue of clear superiority in the technology from which weapons and weapons systems are derived. Today that technology is under relentless attack and the outcomes can be most deadly and decisive for our way of life.
>
> We are meeting this new challenge in major part with a national space program of unparalleled magnitude, cost, and complexity. This program reaches deep into the sciences, and into our industrial strength and military know-how. Its success will be measured in national prestige, scientific advancement, commercial gains, and international security.[27]

In this opening statement, the air force officers writing *The Space Plan* reinforced the idea that there needed to be a close and cooperative relationship between the military and civilian space programs. *The Space Plan* fundamentally reinforced the national space policy presented by President Kennedy before Congress on May 25, 1961, with the exception of the single point that NASA was the primary lead agency. The air force argued in the *Space Plan* that "to arbitrarily separate military from non-military space development responsibility is fundamentally unsound. The capabilities and facilities of the Air Force will be used to support the entire National Space Program, not just the distinctly military portions."[28] This remark completely contradicts

Eisenhower's rationale for the development of NASA, and highlights how displeased the air force had been with the previous structure of the space program. When *Sputnik* was launched into orbit, Eisenhower attacked the Soviet Union for using the "capabilities and facilities" of their military for the benefit of their civilian space program. Now, three years later, air force officers openly advocated civil-military cooperation in space. Believing the nation had a rapidly deteriorating international image, the air force saw the need to combine US space programs to reestablish itself as the world leader in science and technology. In *The Space Plan,* the air force claimed that the Soviet Union's primary purposes in space were to "dominate space and demonstrate Soviet superiority" to the world. By 1961, the opinion that the Soviet Union aimed to control space became an overt institutional position within the air force. Nonetheless, in *The Space Plan,* the air force did not advocate using or basing weapons in space. It did recommend continuing R&D on ASAT and BMD programs, as well as the MIDAS, SAMOS, SAINT, Discoverer, and Dyna-Soar military space programs.[29] Also, the air force again spoke of the use of satellites for mapping, target identification, navigation, missile warning, and meteorological surveillance. Even while concerns escalated over the Soviet Union's attempts to control space, the air force substantially backed away from its 1958 position of arming the heavens, and refocused on the US space program as a way to recapture international prestige and boost technological capabilities. Concerning the evolving air force vision of space, the Air Staff briefed Secretary of the Air Force Eugene Zuckert and the Bioastronautics Group of the president's Science Advisory Committee in October 1961. After the briefing, Secretary Zuckert recommended revisions. Based on Zuckert's revisions, the OSD withheld its official approval. On December 4, 1961, General LeMay authorized the deputy chief of staff for research and development, Lt. Gen. James L. Ferguson, to create a "task force" to prepare a follow-on to *The Space Plan,* and to focus on "outlining the costs and schedules" detailed in *The Space Plan.* The task force prepared fiscal justification and budget projections for the USAF proposed space programs for FY 1963 and FY 1964. In February 1962, Ferguson presented his recommendations to Congress.[30] The task force's report focused on the need to increase the military's space budget and recommended that the United States "exploit space." Ferguson advocated that Congress authorize $1.31 billion dollars for FY 1963 and $1.86 billion dollars for FY 1964. These budget figures reflected a substantial increase over the OSD's budget targets of $826.6 million for FY 1963 and $1.32 billion for FY 1964. Ferguson stressed that the budget increases would enable the United States to "retain military superiority and to insure the peaceful uses of space." Ferguson also suggested that the additional funds could improve the cooperative efforts between NASA and the air force, stating that the two "must be mutually supporting rather than competitive."[31]

In his remarks Ferguson repeated the ideas outlined in AFOS 2/2 and *The Space Plan.* Since Kennedy's inauguration, the air force had lobbied for a more active role in the development of the national space program. Despite remarks by Generals White and Ferguson about the air force desire to work more closely with NASA, air force leaders remained deeply committed to manned space systems such as the dynamic soaring vehicle (Dyna-Soar), which directly overlapped NASA's efforts in Projects Mercury and Gemini. USAF advocacy of Dyna-Soar led to a bitter rivalry with NASA and again fueled public suspicion about the air force desire to control both the civilian and military space programs. The controversy and debate about the utility of Dyna-Soar, which led to its eventual cancelation, serves as an excellent example of how the air force sought to position itself as the lead agency in space science and exploration.

The Controversy

The dynamic soaring vehicle program, Dyna-Soar, had an interesting life prior to Secretary of Defense Robert S. McNamara's termination of the program in December 1963. The program began with a secret 1951 Bell Aircraft study called Project Bomber Missile (BOMI).[32] Bell engineers envisioned BOMI as a reusable launch vehicle functioning as a bomber, manned reconnaissance platform, or passenger craft.[33] BOMI was not the only research program on reusable boost-glide vehicles in the early 1950s. Programs such as the high-altitude reconnaissance craft (118P), long-range reconnaissance vehicle (Brass Bell), and the rocket bomber (ROBO) encompassed roles similar to those of BOMI. Seeing the strategic promise in BOMI, ROBO, and Brass Bell, the air force expressed interest in the concepts. Prior to October 4, 1957, the air force had limited R&D funds for reusable launch vehicles; however, the launch of *Sputnik* provoked Eisenhower to get an increase from Congress for the air force's R&D budget for space projects. In November 1957, the air force consolidated the various boost-glide programs into a single program and authorized a developmental plan.[34] By March 1958, the air force had selected Boeing as prime contractor for the Dyna-Soar program.

Artist's rendition of Dyna-Soar, 1957. NASA, courtesy of nasaimages .org

After completing initial feasibility and design studies, Boeing proposed military missions for the vehicle to the air force. Boeing engineers identified "global surveillance and orbital command post" as the two main missions for Dyna-Soar.[35] These two missions differed greatly from the "rocket bomber" ideas considered at the beginning of the decade. Despite the optimism of Boeing's engineers, Eisenhower by 1958 openly advocated for the peaceful use of space, and therefore the idea of using Dyna-Soar as a space or orbital bomber fell out of political favor. In the post-*Sputnik* years, Dyna-Soar had to compete for budget allocations with military satellite systems and NASA manned space missions.

While the air force worked to complete the initial phase of the Dyna-Soar project, NASA had already begun Project Mercury and moved quickly toward starting Project Gemini. Having lost the bomber role for Dyna-Soar, the air force expanded the scope of the Dyna-Soar project beyond the missions identified by Boeing engineers. The air force supported Dyna-Soar as a research vehicle for collecting data on manned reentry and maneuverability in space. NASA argued that the Mercury and Gemini programs already provided research platforms designed to collect similar data. With cumulative costs in excess of $800 million, the Dyna-Soar project lost its appeal to Eisenhower. By the end of Eisenhower's presidency, the future of the USAF Dyna-Soar was highly questionable because it duplicated the research program outlined by NASA programs.

With the change in presidential administrations in 1961, the status of the Dyna-Soar program failed to improve. Secretary of Defense Robert S. McNamara told the air force to justify Dyna-Soar "on the grounds of an identifiable military mission."[36] The air force responded by stating that Dyna-Soar could be used for "reconnaissance, rescue, satellite inspector, bomber, and ferry" missions.[37] Because the United States already had satellite programs designed for reconnaissance and inspection, and since Kennedy rejected the weaponization of space, the air force's redundant military justifications for Dyna-Soar flopped.

Dyna-Soar's escalating costs and duplication of NASA's research objectives in Project Gemini concerned McNamara. He placed Dyna-Soar "under review," thinking that the Air Force had failed to provide adequate military justification for the expensive program.[38]

Despite McNamara's skepticism, Dyna-Soar retained support in Congress. On April 17, 1962, Wisconsin Congressman Melvin R. Laird stated: "For all our hopes that space may not become an arena of future conflict, we must clearly recognize that if man can go into space for peaceful exploration and research, he can use this same environment for military purposes. Those people in this country that are today placing all emphasis on getting to the moon at the cost of billions of dollars are making a mistake. I am convinced that there will be future military weapons systems operating in space and some, perhaps most, of these systems will have to include men to be most effective."[39] Laird, citing Soviet generals' comments about the importance of space for future wars, maintained that the Dyna-Soar program remained a vital element in countering the Soviet Union's presumptive designs to control space. In June 1962, with support from Congress, the air force re-designated the Dyna-Soar program X-20, as a way to reinforce the research objectives of the program.[40]

The use of the X designation signaled the experimental nature of the program and increased cooperation with NASA, as with the X-15. On September 20, 1962, Under Secretary of the Air Force Joseph V. Charyk spoke at the unveiling of a mockup of the X-20 in Las Vegas, Nevada, before the Air Force Association's annual convention. After providing highlights of the successful joint Air Force–NASA X-15 program, Charyk called the X-20 the next step in cooperative aerospace research.[41] Charyk also affirmed the air force position: viewing the X-20 as a platform for researching manned exit and reentry from space. While highlighting the research value of the X-20 program, Charyk shied away from the military potential of the vehicle, claiming, "[T]he X-20 does not represent a vehicle for a specific military job." By late 1962, the air force still pitched Dyna-Soar/X-20 as a research program, but failed to specify military missions for the craft.

Despite the air force's claimed research potential of the X-20, Secretary of Defense McNamara remained unconvinced. On December 10, 1963, he publicly announced the cancellation of the X-20 program, stating that the X-20 did not meet the "current requirements for a program aimed directly at the question of man's utility in space."[42] Although the air force lost one major space program, McNamara during the same news conference announced the creation of a new space program for the Air Force, the Manned Orbital Laboratory (MOL).

The MOL replaced the X-20 and provided the Department of Defense and the air force with a program to "determine the military usefulness of a man in space," at a cost $100 million less than that of the terminated Dyna-Soar program.[43] The new program fit within the space research agenda of the DoD, not entailing the use of weapons in space and not infringing on NASA's Gemini program, but unfortunately it placed the air force in a position equal with NASA, not in the lead role for which it had lobbied. In fact, MOL typified Kennedy's national space program by using a "modified Gemini capsule" and being closely connected to the work being done by NASA in its manned space missions. In MOL, the administration saw the civilian and military space programs working in concert.[44]

Kennedy's support for MOL demonstrated his commitment to the national space programs and his rejection of USAF attempt to recast the program under military direction. MOL offered a great opportunity for the president and the secretary of defense to streamline the space programs. Kennedy and McNamara used NASA's Gemini program as a foundation to assist the air force with its interest in manned spaceflight. The air force support allowed Kennedy and McNamara to appease USAF interests in space, while also emphasizing the US position on the peaceful use of space. Having appeased the

Artist's rendition of Manned Orbital Laboratory (MOL), 1960. NASA, courtesy of nasaimages .org

air force and expanded NASA's space program, Kennedy focused on keeping weapons out of space.

Moving toward Disarmament

To improve the international image of the United States, Kennedy strove to revise his space policy. Starting early in 1961, Kennedy's science advisor, Jerome Wiesner, recognized the advantages gained by the Soviet Union in the aftermath of *Sputnik.* Wiesner had pointed out in a memorandum dated February 20, 1960, that the Soviet Union had a three-to-one advantage in lift capability, which gave the Soviet Union several advantages for space missions and in the development of strategic rockets.[45] Wiesner stressed that the United States could expect a three-year lag before reaching parity with the Soviet Union's lifting capability. While the United States worked to reach parity, Wiesner forecast "continued embarrassment" for the United States.

To provide some positive information for Kennedy, Wiesner observed that the United States maintained a distinct advantage in exploiting the "scientific purposes" of space. According to Wiesner, the United States had achieved many firsts, such as "the definition of the Van Allen belt; the precise geodetic use of an artificial satellite to obtain refined information on the size and shape of the earth; the achievement of both active and passive communication satellites; discovery of a large electrical current system about the earth; successful use of weather satellites; the first measurement of interplanetary magnetic fields; radio communication at interplanetary distances; and the first simultaneous observation of solar disturbances and associated magnetic storms from interplanetary space and on earth."[46] Although he acknowledged the importance of these accomplishments, Wiesner also noted: "Unfortunately, it is more difficult to dramatize these things than it is the massive performance by the Soviet Union." To combat the Soviet Union's prestige, Wiesner recommended readjusting the US space program.

Beginning in 1961, Kennedy moved away from the space policy that Eisenhower and his National Security Council outlined in NSC-5814. Kennedy advocated "four basic principles" necessary to reshape the US space program.[47] Kennedy believed that "scientific, commercial/civilian, military, and national prestige" objectives needed to drive the program. According to the NSC, the United States only lagged in programs designed to "enhance national prestige."[48]

Based on Wiesner's analysis, the NSC believed the United States had already achieved success in the "scientific and military" fields. Military reconnaissance and communication satellites provided two such areas in which the United States led the Soviet Union. However, these programs remained highly classified and therefore the administration could not use them to bolster the prestige of the United States. Therefore, despite Kennedy's belief that the United States needed to move beyond the policies of the Eisenhower administration, he accepted the MIDAS, SAMOS, and Advent programs because they provided the United States with advanced technological capabilities, yet he allowed the public to believe that he was reinvesting in the civilian uses of space.

Based on analysis by Wiesner and the NSC, Kennedy's support of a large-scale effort to land and return men from the surface of the moon provided the United States with a high-profile program that could seize the attention of the world. Kennedy also supported the manned lunar mission so that the United States could rectify deficiencies in heavy-lift rocket boosters that Wiesner had recognized.

Furthermore, Kennedy believed that the United States needed to continue to work toward keeping space free of weapons.[49] On September 25, 1961, Kennedy outlined his position in an address before the United Nations. Kennedy proclaimed that "the new horizons of outer space must not be driven by the old bitter concepts of imperialism and sovereign claims."[50] He continued: "The cold reaches of the universe must not become the new arena of an even colder war." In this speech, Kennedy heralded the US commitment to keeping space free of weapons.

To advance the US agenda on the peaceful use of space, Kennedy exchanged a series of letters with Khrushchev discussing the possibility of bilateral cooperation in space.[51] Although the two nations had briefly discussed a joint space mission after *Sputnik,* Eisenhower and others believed that a joint program would provide the Soviet Union with an opportunity to gather intelligence on US electronic components.[52] Eisenhower rejected the idea. Kennedy, however, viewed it as a way to demonstrate the peaceful intentions of the United States. Kennedy and Khrushchev exchanged letters reiterating the need to work together to keep space weapons-free.[53] Both men also acknowledged the need to focus on the "peaceful purposes of space."[54]

Working toward banning weapons from space, Kennedy, in the 1962 Geneva Summit with Khrushchev, proposed banning nuclear weapons tests in outer space. In support of Kennedy's proposal, in a speech before San Francisco's Chamber of Commerce in May 1962, Deputy Secretary of Defense Roswell Gilpatric publicly presented the military space programs of the United States as being within the president's "national objective of using space for peaceful purposes."[55]

Gilpatric did not reject the militarization of space. Rather, he objected to the use of offensive weapons in space. This distinction allowed Kennedy to work toward an international agreement to keep space free of weapons, while also supporting the development of ground-based ASAT and BMD systems for the United States.

In Monterey, California, on May 2, 1962, Gilpatric had stated that the United States had not established or designed programs for offensive capabilities in space, although he did acknowledge US work on "defensive systems."[56] In the San Francisco speech before the San Francisco Chamber of Commerce, Gilpatric warned that the Soviet Union would "not cooperate" in an international treaty to ban weapons from space, adding that the United States needed to be prepared to "neutralize" Soviet attempts to control space.[57]

Gilpatric's comments and the administration's idea about the use of "defensive" weapons in space meshed with SAINT, the satellite inspector program. In the aftermath of *Sputnik,* the air force designed SAINT, envisioned as an inspector satellite with an additional capability to "kill" hostile satellites. While SAINT was conceived as a weapons system, Gilpatric and others in Kennedy's administration believed that a reoriented SAINT program could provide the United States with a defensive capability to protect the world from Soviet domination of space. The potential "offensive" use of SAINT never troubled Kennedy and his administration. Kennedy believed that the Soviet Union's Fractional Orbital Bombardment Systems (FOBS) remained fundamentally different from SAINT. Throughout the summer of 1962, Kennedy maintained his commitment to banning offensive weapons from space. Officially, the administration said:

> The DoD has for a long time conducted research on possible military uses of space. This has been publicly known and is not inconsistent with either (1) our effort through disarmament to preclude use of space for aggressive purposes or (2) our efforts to encourage and support the cooperative use of space for peaceful purposes."[58]

Using this policy position, Kennedy authorized the National Aeronautics and Space Council (NASC), chaired by Vice President Lyndon B. Johnson, to develop a new US policy on outer space. The members of NASC identified the US space programs as "basic to the country's world leadership responsibilities, to its security and to its economic, scientific, and cultural growth."[59] NASC also maintained that the United States needed "to explore and use outer space to further national progress, security, well-being, and to encourage international cooperation in the interest of world peace and for the benefit of mankind."

After establishing the objectives of the US space program, NASC classified its missions as scientific, technological, military, and diplomatic. NASC made sure the report remained general enough to cover both the civilian and military space programs and emphasized a single national space program, not two separate programs.

The public acknowledgment by Kennedy and the members of NASC that the civilian and military space programs of the United States worked in concert with one another to achieve a national goal was a major break from Eisenhower's public rhetoric that the two programs needed to be separated.[60]

Kennedy voiced the same basic message about the peaceful use of space as had Eisenhower. However, Kennedy also strove to present the military and civilian space programs as a single national space program. For Kennedy, the program encompassed the gamut of civilian and military objectives and missions. He stressed that both elements of the national space program remained vital to his objective of keeping space free of weapons.[61] By acknowledging the importance and necessity of military satellites in preserving peace, Kennedy openly justified to the international community America's continued militarization of space as a way to preserve peace and stability. By August 1962, the administration focused on completing an arms control agreement on space weapons. On August 9, 1962, adviser Jerome Wiesner sent a memorandum to President Kennedy, titled "Banning of Weapons of Mass Destruction in Outer Space."[62] In his memorandum Wiesner stated: "There is no question that it would be technically feasible to design a variety of weapon systems employing nuclear weapons in space. These weapon systems could be designed either for target bombardment with accuracies approaching those obtained with ballistic missiles or for the detonation of extremely high yield warheads—possibly as large as 1000 Megatons—directly in orbit."[63] Wiesner listed three arguments supporting space weapons. First, "the existence of space based weapons systems would add to international technical-military prestige" and produce a psychological benefit for the possessor. Second, the "stationing of weapons of mass destruction in orbit" could allow the United States to disperse weapons and "avoid the destruction associated with a counter-force attack." And third, the use of space-based bombardment satellites would reduce the risk of a surprise first strike.[64] Having established the basic arguments for space weapons, Wiesner recommended that the United States pursue a more pacific and altruistic path which was more consistent with its stated space policy. While mentioning the USAF lack of interest in "space-based nuclear weapons systems," Wiesner stressed that the Soviet Union might still decide to deploy a space-based system to gain a psychological benefit and improve its strategic position. After issuing his warning about the Soviet Union, Wiesner recommended that Kennedy proceed with a proposal to ban weapons of mass destruction in space, saying, "It could serve as a tactic to initiate some activity in the field of disarmament and could prevent a very expensive new dimension to the arms race with its unpredictable psychological reaction which might affect the overall level of military activities of both sides."[65] Wiesner's analysis provided Kennedy with a foundation to begin constructing an international proposal committing the United States to banning weapons from space. Wiesner's speculation about the Soviet Union's interest in proceeding to base weapons in space heightened Kennedy's concerns about the deployment of FOBS by the Soviet Union. As the summer of 1962 faded into the fall, Kennedy worked toward banning weapons from space, while his concern over the Soviet Union's research and development of FOBS intensified.

On September 15, 1962, the director of the US Arms Control and Disarmament Agency (ACDA), William C. Foster, presented a draft of the agency's *Recommendations Respecting U.S. Approach to a Separate Arms Control Measure for Outer Space.* In this memorandum Foster emphasized, "[I]t is increasingly clear that questions related to the use of outer space are becoming a major international issue."[66] Paralleling the stance advocated by Wiesner, Foster stated: "ACDA's view is that the United States' position in meeting probable international pressure would be strengthened by a well-defined position and an affirmative approach toward a separate measure to prohibit the placing in orbit of weapons of mass destruction."[67] Beyond his endorsement of an arms control treaty, Foster advocated that Kennedy consider "reliance on a unilateral statement of intention" and "expression of willingness on the part of the United States to negotiate a more formal arrangement." Foster's recommendations supported Kennedy's agenda to preserve space as a weapons-free environment and furthermore provided the administration with a political tactic by which they could counter concern about the Soviet Union's interest in weaponizing space. While Kennedy advanced his agenda, he had to be careful not to subvert a movement within the United Nations to ban all military uses of space, which Kennedy did not endorse.[68] Like Eisenhower he believed in the use of space for specific non-aggressive military missions (reconnaissance, weather, communication, and navigation). A complete ban on all military uses of space was therefore out of the question.

The Soviet Union challenged Kennedy by pushing for an outright ban on all military uses of space. The Soviet delegation to the United Nations proposed a resolution to the Outer Space Council "to prohibit the use of space for military intelligence purposes."[69] In response, Kennedy defended the US military satellite programs as necessary due to the secret and closed nature of the Soviet Union. Facing international pressure on the issue of the militarization of space, Kennedy advocated that the United States maintain its position against the use of weapons in space, yet maintain its support for developing the non-aggressive military use of space.

Spearheading the US-led international effort to ban weapons from space, Deputy Under Secretary of State U. Alexis Johnson, on September 21, 1962, circulated a memorandum to the UN General Assembly to provide general guidance on the issue of space disarmament.[70] Johnson attached the following position papers to his memorandum: "U.S. Outer Space Policy in the UN General Assembly," "Definition of the Peaceful Uses of Outer Space,"

"Photographic and Observation Satellites," "United States Military Space Programs," and "Effects of High Altitude Nuclear Explosions." Together the policy papers outlined the administration's support for non-aggressive military satellite programs, while also focusing on the use of space for peaceful purposes.[71]

Despite the strong stance taken by the United States, Deputy Under Secretary of State Johnson believed the United States faced increased "challenges to their space programs."[72] Johnson stressed that the international community's "focus would probably be on "alleged U.S. military (weapons) ambitions in space, reconnaissance satellite program, space experiments which affect the use of space by other countries, and the U.S. space communication program."[73]

To deflect international criticism of America's military space programs, Johnson, in "U.S. Outer Space Policy in the U.N. General Assembly," recommended five positions to Kennedy in countering the accusations by the Soviet Union, the first being that Kennedy reinforce his pledge to preserve open access to space for all nations. Second, Johnson recommended that Kennedy emphasize the lack of US programs "to place weapons of mass destruction in space" and stress US leadership in the drive to ban weapons from space, to counteract international accusations. Third, Johnson advised Kennedy to accentuate the civilian benefits (weather forecasting, research surveys, mapping, and geodesy) of "observation satellites," as well as the openness that the United States maintained in its development of "photographic and observation satellites." Johnson observed that international law did not prohibit the use of these non-aggressive satellite systems. Fourth, Johnson advised Kennedy to avoid criticism over US nuclear testing by reminding the international community of America's open notifications of its nuclear experiments and that the president's decision to conduct experiments in space did not infringe upon his commitment to maintain the openness of space.[74] Last, Johnson recommended that Kennedy allow the commercial control of space communication assets to demonstrate the US interest in developing a "global and non-discriminatory system of satellite communication with international participation."[75] Kennedy used Johnson's advice to solidify his commitment to recast the United States as the world leader in the peaceful uses of space.

Having established the US position on space vis-à-vis Soviet rhetoric, the State Department cleared up existing confusion over the phrase "peaceful uses of outer space." In a policy paper attached to his September 21, 1962 memorandum, Deputy Under Secretary of State Johnson described the "widespread confusion over the distinction between peaceful and aggressive, and military and civilian" uses of space.[76] To remedy the confusion, Johnson advised Kennedy to clarify the US position by asserting that "peaceful" and "aggressive" uses of space were not synonyms for "civilian" and "military" uses. Johnson held that the Soviet Union's characterization of the uses of space as "peaceful/civilian" and "aggressive/military" was wrong. U. A. Johnson provided Kennedy with policy guidance to support the president's decision to focus on developing a national space program that rested on the underlying principal that the United States promoted the peaceful uses of space. In many ways, Johnson's memorandum captured the administration's two years of work at further militarizing space while also keeping space free of weapons. As intelligence about the Soviet Union's FOBS program increased in late 1962 and early 1963, Kennedy bolstered his commitment to banning weapons from space.

FOBS and Program 437

As the potential threat of FOBS emerged, Kennedy worked to ensure that the United States had the capability to protect itself. In December 1962, the RAND Corporation's J. M. Goldsen completed *Military Space Activities and the Cold War* in which he summarized the Soviet Union's military space capabilities.[77]

Although he recommended "negotiations and even cooperation with the Russians in space," Goldsen believed that the United States might "find it necessary to pursue military action in space." Goldsen based his recommendation on the analysis of statements of, writings by, and interviews with key Soviet officials such as Nikita Khrushchev, Defense Minister Rodion Malinovsky, and former chief of the general staff Marshal Vassily Sokolovsky. Throughout his report, Goldsen highlighted phrases, passages, and speeches by Soviet officials who stressed the strategic importance of space.

Specifically, he included a passage from Sokolovsky's *Military Strategy* emphasizing that the Soviet Union should obtain the capability to "offset the imperialists' [efforts in space]."[78] According to Goldsen, the rhetoric of Khrushchev, Malinovsky, and Sokolovsky pointed toward the Soviet Union's transition from the scientific and peaceful uses of space toward more bellicose objectives. Goldsen indicated that "bombardment satellites and anti-satellite systems" were emerging as the major military space systems in the Soviet Union's quest to control space.

The idea of using satellites for bombardment was nothing new. In fact, RAND engineers in 1946 had contemplated the idea before quickly dismissing it.[79] In the late 1950s, as missile and rocket technology advanced, the technological capability of using satellites as bombardment platforms received greater attention. The air force even temporarily considered using satellites as bombardment platforms before Eisenhower rejected this in favor of maintaining space as a peaceful frontier. The Soviet Union, however, continued to consider, refine, and develop the idea throughout the 1950s and 1960s.

For Kennedy, the Soviet Union's interest in bombardment satellites proved troubling. Throughout 1963, Kennedy remained deeply concerned over the Soviet Union's apparent willingness to orbit them.[80] In February 1963, U. Alexis Johnson issued a memorandum outlining the administration's position in response to the Soviet Union's desire to orbit nuclear weapons in space.[81] Parallel to what Jerome Wiesner told Kennedy in his August 9, 1962 memorandum, Johnson recommended that Kennedy stress the "military insignificance" of basing nuclear weapons in space. Johnson also advised Kennedy to reaffirm his public commitment to disarming space, even if the Soviet Union proceeded with FOBS.[82] In the event the Soviet Union deployed FOBS, Johnson outlined four options open to the president:

- To declare the Soviet action a threat to the peace and (a) to present the Soviets with an ultimatum to remove their space weapons on threat of our shooting them down—or (b) to obtain a U.N. General Assembly resolution calling for the Soviets to remove their weapons from space—and (c) if necessary, in either case the U.S. would shoot them down.
- To declare the Soviet action a threat to the peace compelling the U.S. and other nations to reappraise Soviet intentions, and to indicate the strength of our objection by a range of strong diplomatic actions short of a general ultimatum and not involving shooting down their satellites.
- To place weapons in space ourselves, stressing that our reluctant decision to deploy weapons in space was compelled by the Soviet action.
- To attempt to deflate the Soviet action by playing down its military significance and castigating the Soviets for a dangerous propagandistic action.[83]

Judging the options presented by Johnson, option three was out of the question based upon Kennedy's commitment to the peaceful use of space. Options one and two remained available, but inconsistent with the president's pacific position. Option four was consistent with the administration's position and devoid of the "aggressive" elements found in options one and two. The Department of State worked through spring 1963 to develop planned statements in preparation for the Soviet Union's deployment of FOBS, in an effort to provide the administration with diplomatic rhetoric to counter the Soviet Union.

In May 1963, the administration's concern over FOBS intensified.[84] The US ambassador to Czechoslovakia, Outerbridge Horesy, officially objected to the Soviet Union exhibit on "Man and Flight into Space," sponsored by the Soviet Academy of Science, the USSR government, and the Czechoslovak Ministry of Education and Culture, which displayed "diagrams demonstrating the effectiveness of the Soviet Union's 'global missile' and anti-missile system." Ambassador Horesy protested that the elements in the diagrams demonstrating the Soviet Union's use of fractional orbiting satellites to attack the United States from a southern trajectory, effectively bypassing America's early warning networks, constituted an inappropriate threat to international strategic stability by depicting the "aggressive" Soviet intentions.

Ambassador Horesy also protested the presentation's depiction of FOBS, claiming that it contradicted the official policy of the Czech government to use space for peaceful purposes. The displays in Czechoslovakia provided further evidence for Kennedy of the Soviet Union's interest in orbital weapons. Kennedy believed that this situation showed that the United States needed to achieve its goal of banning weapons from space and seemed an excellent opportunity for the United States to once again demonstrate to the international community its pacific intentions toward space. Conversely, the Kennedy administration also capitalized on this event by highlighting the Soviet Union's interest in weaponizing space for national security purposes.

By October 1963, ACDA produced a proposal outlining an arrangement between the Soviet Union and the United States on the use of nuclear weapons in space. An attachment to the memorandum dated October 1, 1963, titled "A U.S.-Soviet Arrangement Concerning the Placing in Orbit of Weapons of Mass Destruction," recommended "a parallel unilateral statement of intention by the two countries. A joint statement of intentions by the two countries. A resolution of the General Assembly. A resolution of the Eighteen-Nation Committee on Disarmament. Or an executive agreement or treaty."[85] The ACDA suggested that Kennedy pursue a bilateral agreement between the United States and the Soviet Union concerning the use of weapons in space as a way to build additional interest in the General Assembly of the United Nations.[86] Kennedy recognized the need for the United States to neutralize attempts by the Soviet Union to orbit bombardment satellites. Working to produce an international or even bilateral disarmament agreement, Kennedy strove to improve the ASAT capabilities of the United States, as a military hedge against the possibility that the Soviet Union might go ahead and develop FOBS.

US interest in ASAT issues emerged prior to Kennedy's concerns about the Soviet Union's orbital bombardment satellite in the 1960s. During the late 1950s, air force officers, as well as key individuals in Eisenhower's cabinet, believed that the United States needed the capability to destroy hostile satellites. In the years after the launch of *Sputnik,* anti-satellite capabilities became a major concern for the US military. As noted earlier, even as Eisenhower voiced concern over the weaponization of space, he did agree to allow research into ASAT capability. SAINT, conceived as an inspector and potential killer of hostile satellites, remained the major ASAT program for the United States.

However, in the transition between the Eisenhower and Kennedy administrations, the SAINT program evolved into a satellite inspection system rather than a weapons system. Technical setbacks and cost overruns plagued the SAINT program as Kennedy took office in 1961 and continued afterward. As a result Secretary of Defense McNamara canceled the program in December 1962.[87] As Kennedy's concerns over the Soviet Union's orbital bombardment satellite mounted throughout 1961 and 1962, the need for a replacement ASAT system evolved. The US nuclear testing program provided research potentially useful for development of an ASAT system.

Since 1960, the Atomic Energy Commission (AEC) and the DoD had been conducting the Fishbowl series of tests to further understand the effects of high-yield nuclear detonations at high altitude. In the Starfish Prime test on July 9, 1962, the United States detonated a 1.4-megaton nuclear device at an altitude of roughly 248 miles. This test produced a significant electromagnetic pulse (EMP). The EMP "seriously damaged the solar panels of three satellites" and the prompt radiation released from the blast exposed four other satellites to high levels of radiation, resulting in mechanical decay.[88] The information gathered on high-yield and high-altitude nuclear tests by AEC and DoD confirmed the potential of EMP to damage and destroy satellites.[89]

As a result of the findings in the Starfish Prime test, Gen. Bernard Schriever, USAF, proposed a new ASAT system to Secretary of the Air Force

Eugene Zuckert on September 12, 1962: ground-based Thor ICBMs armed with nuclear warheads. Zuckert quickly approved the idea. Secretary of Defense McNamara concurred with Zuckert and authorized the program on November 20. In December, the air force began work on the new ground-based system, known as Program 437.

In the summer of 1963, Program 437 was a national priority for Kennedy.[90] While Zuckert, McNamara, and Kennedy supported the development of the program, the army and navy proposed rival ASAT systems. The army recommended upgrading its Nike-Zeus ABM system for the ASAT role, while the navy proposed Early Spring, an upgrade of the new Polaris SLBM.[91] The air force fended off the army and navy by arguing that Nike-Zeus lacked an "inspection component" vital to the ASAT mission and that the navy's sea-based proposal "offered no unique advantages" over the ground-based ASAT system.[92] Despite the air force attempt to bolster its ASAT program, McNamara gave the army approval to develop the Nike-Zeus ASAT, which came to be known as Program 505. While the Nike-Zeus satellite interceptor lacked the range and payload lift capability of the air force's Thor, the army's ABM missile proved itself in a series of successful space interceptor tests throughout the summer of 1963.[93]

The successful tests and the advanced development of Program 505 led McNamara to order on June 27, 1963, a "single Nike-Zeus missile to stand ready to intercept a Russian satellite." In fielding this ASAT system, the Kennedy administration succeeded in providing the United States with the initial operational capability (IOC) to down hostile satellites. The air force continued development of Program 437, which replaced the army's Program 505 in 1964 as the nation's ASAT system.[94]

From Kennedy's perspective, Programs 505 and 437 did not violate his pledge to keep space free of weapons. Rather, Kennedy and his administration justified the ground-based nuclear ASAT programs as defensive and non-aggressive weapons necessary to guard against the Soviet Union's designs to control and dominate space. Against heightened concerns about the Soviet Union's interest in developing orbital bombardment satellites, Program 505 and Program 437 provided Kennedy with a capability to counter the Soviet Union's FOBS. The ASAT programs begun under Kennedy allowed the United States to reinforce its commitment to the peaceful use of space, by threatening to use force if necessary. From the perspective of the Kennedy administration, the initial operational capability of US ground-based anti-satellite systems further cemented the US commitment to stand firmly behind its pledge to maintain space as a frontier free of weapons.

Conclusion

On November 22, 1963 in Dallas, the Kennedy era ended abruptly. Despite the brevity of his administration, John F. Kennedy had profoundly influenced the national space policy of the United States. With NASA, Kennedy had seen an opportunity to reestablish the United States as the world leader in science and technology. In a drastic break with Eisenhower's policy of not overtly challenging the Soviet Union in "space stunts," Kennedy announced his idea to send a manned mission to the moon. Kennedy believed that this challenge embodied the elements the United States needed to reestablish itself as the world leader in space technology. Further, he wanted to reestablish America's technological prestige with heavy-lift booster rockets, a technology in which the Soviet Union possessed advantage.

Although emphasizing the civilian space program, Kennedy committed the nation to the further development of non-aggressive military uses of space. He and Secretary of Defense McNamara embraced the majority of the military space systems funded under Eisenhower. While his administration cancelled the costly and troublesome Dyna-Soar and SAINT programs, Kennedy recognized the importance of the Discovery/Corona, MIDAS, Advent, Transit, TIROS, and Nimbus systems. Photoreconnaissance, early warning, intelligence gathering, communication, navigation, and weather data collection remained vital to the national security strategy of the United States, and Kennedy embraced the continued use of military satellites, while also recognizing the potential civilian uses of these space-based systems.

The dual-use capability of many of the above-mentioned satellite systems allowed Kennedy to assert the US commitment to the peaceful use of space, although international concerns increased over its militarization. Kennedy maintained that the non-aggressive military programs of the United States remained necessary for the preservation of peace and stability. He also pointed out the potential civilian benefit of communication, weather data collection, and navigation satellite systems.

In an effort to further his administration's commitment to the peaceful use of space, Kennedy began negotiations with the Soviet Union and continued working with the United Nations to ban weapons from space. Kennedy's decision to strive for a space arms control agreement coincided with the Soviet Union's interest in FOBS. Concerned by the bellicose rhetoric of political and military officials from the Soviet Union and their intentions to use orbital bombardment satellites, Kennedy responded to the FOBS threat in two distinct ways, both of which reaffirmed his pledge to keep space a peaceful frontier.

First, the ACDA outlined a public policy position for Kennedy to counter the Soviet space threat by spearheading a drive to ban weapons from space. By employing this tactic, Kennedy reinforced his commitment to the peaceful use of space within the international community and to establishing the United States as the lead nation in this endeavor. Recommending a framework for keeping arms out of space also allowed Kennedy to combat global criticism of the US military space program. Leading the drive for an arms control agreement on space, Kennedy improved the international image of the United States, while simultaneously casting the Soviet Union as a barrier to global peace.

Second, Kennedy ensured the United States had an anti-satellite weapons system capable of countering a FOBS threat from the Soviet Union. In an effort to remain committed to the administration's public policy of disarmament, McNamara chose ground-based, nuclear-tipped ASAT interceptors over space-based systems.[95] Program 505 and Program 437 provided the United States with an operational capability to intercept hostile satellites. Kennedy and his national security staff viewed ASAT weapons systems as defensive in design and rationalized that ASAT development supported the administration's position on space. As president, Kennedy maintained Eisenhower's pledge to pursue only the peaceful use of space. Kennedy advanced his ideal by formalizing the objective within the framework of the United Nations and in bilateral negotiations with Khrushchev. While striving to achieve peace, Kennedy supported the continued development of military space systems and the creation of ground-based ASAT systems. Using a dual approach, Kennedy sought to convince the world of America's noble intentions, while also ensuring that the United States maintained the ability to fight for the peaceful use of space.

Chapter 4: Lyndon Johnson and Space as a Weapons-Free Frontier, 1963–1967

Under the leadership of Lyndon B. Johnson, the United States solidified its commitment to the military use of space and at the same time presided over the fulfillment of the objective of securing space as a weapons-free frontier by signing the Outer Space Treaty (OST) in 1967. Johnson also furthered US commitments to the use of ground-based ASAT and BMD systems to maintain national security. Like his immediate predecessors, Johnson believed in having a national space program that was comprehensive—military and civilian together—to protect space from becoming an armed arena in the Cold War.

Although he signed the OST, Johnson remained concerned about the Soviet Union's research into orbital bombardment systems. Though the OST codified the decade-long quest to keep weapons out of space, the Soviet Union remained interested in developing FOBS. To counter the potential of a Soviet FOBS program, Johnson enhanced the ASAT capability of the United States. Building upon the space policy of Eisenhower and Kennedy, Johnson believed the United States needed to enforce the OST. As with Kennedy, Johnson recognized that ground-based ASAT and BDM systems used space intercepts, but he believed that this did not violate the OST because the interceptors were not stationed in space. In Johnson's view, space remained a weapons-free environment.

While technically both ASAT and BMD systems provided the United States with space weapons, the ground-based systems differed greatly from those conceived in the late 1950s that would have based weapons or warheads in space or on the moon to achieve the same purposes. Although Johnson rejected the notion of weaponizing space in this way, he accepted the need for defensive systems. By 1967, space had become a vital military medium and Johnson sought to protect it for the national security of the United States and the scientific benefit of the international community.

Johnson balanced the military and civilian space efforts of the United

States. Unlike Eisenhower, Johnson publicly supported a "national space program" that encompassed both the military and civilian programs. While maintaining the same basic military missions envisioned for satellites by the RAND Corporation in 1946, Johnson used NASA's civilian missions to reinforce the US commitment to using space for peaceful purposes. During Johnson's presidency, the international community benefited greatly from the use of space for navigation, mapping, communication, and weather data collection through Johnson's opening of NASA's TIROS, Transit, and Nimbus satellite programs to the public. Johnson's national space program improved the international image of the United States, while also strengthening the US military space program.[1]

In many ways, Johnson's tenure served as a capstone completing the initial phase of the space age. Under his direction, the United States maintained an active and hardy military space program designed to provide intelligence, reconnaissance, communication, and aerospace research deemed necessary during the Cold War. Many of the programs first envisioned in the days immediately after the launch of *Sputnik* became vital assets in the Cold War struggle against the Soviet Union and its allies.

By the end of Johnson's tenure, the United States had succeeded in reversing its less than stellar international image by using its space program demonstrate superior scientific, technological, and engineering capabilities. By 1968, America had become the world's premier space power.

Senator Johnson and Space

Like John F. Kennedy, Johnson had legislative experience in space issues prior to becoming president—only more so. As Senate majority leader and then vice president, Johnson had worked on space issues. Having presided over the Armed Services Committee, the Satellite and Missile Programs Subcommittee, and hearings covering military and industrial preparedness in the wake of *Sputnik I* and *II,* Johnson had amassed a wealth of knowledge about the US space program and international space issues. As vice president, Johnson became chairman of the National Aeronautics and Space Council overseeing space programs and the research and development of NASA projects, as well as military space programs. By 1963, Johnson had evolved into one of the nation's leading space experts. His background in space and military matters proved vital as he assumed the presidency in November 1963. By this time,

NASA Administrator T. Keith Glennan (left) presents a piece of Mylar from the Echo I satellite to Sen. Lyndon B. Johnson, chairman of the Senate Aeronautical and Space Sciences Committee, August 12, 1960. NASA, courtesy of nasaimages.org

Johnson had already solidified his thoughts on space based on his congressional experiences in the late 1950s. He used his knowledge to ensure that the United States became the world's leading space power. Prior to the Soviet launch of *Sputnik* in 1957, Johnson's experience with space had been minimal. As a member of the Senate Committee on Armed Services, however, Johnson had been concerned about US preparedness in the Cold War. As noted earlier, in the late 1940s and early 1950s, Truman and Eisenhower had been concerned by the military development and increased political influence of the Soviet Union. As a senator, Johnson had been well acquainted with shortcomings in preparedness, specifically "lagging production in aircraft, tanks, and electronics."[2] During the Cold War arms race, President Johnson, like Truman and Eisenhower, confronted the complex issue of national security, and space programs became critical as the arms race evolved. During the first years of Eisenhower's presidency, Johnson, as chairman of the Satellite and Missile Programs Subcommittee, heard testimony from the DCI, Allen Dulles, and the director of the Office of Scientific Intelligence of the CIA,

Herbert Scoville, on the status of the Soviet Union's missile programs. Dulles and Scoville raised three important points concerning progress of the USSR. First, the Soviet Union had a "coordinated native missile and research development program." Second, it had made considerable progress in the development and testing of "short, medium, and intercontinental range missiles" and in the design of nuclear warheads. Third, based upon rapid progress in the Soviet Union's missile development, Dulles and Scoville testified that by 1957 the Soviet Union "would have the capability to orbit earth satellites, using boosters that used the same technology developed for their missiles."[3]

Dulles and Scoville emphasized the strategic balance between the USSR and US missile programs, but questions about the use of space and satellites crept into the discussion. Throughout the first decade of the Cold War, RAND, the air force, and Eisenhower recognized that space and satellite programs significantly contributed to the prestige of the country. Members of Eisenhower's cabinet firmly believed that the United States would garner benefits from pursuing these technological programs. As Dulles and Scoville predicted, the Soviet Union launched the first man-made satellite, in October 1957, capturing the world's imagination. A month later, the Soviet Union launched *Sputnik II,* and the United States seemed to lag even further behind the Soviet Union in the space race.[4]

In the wake of the USSR successes in space, Congress held hearings on the status of the US satellite program and the impact of *Sputnik I* and *II.* At the opening of the hearings, Senator Johnson stated, "[T]here were no Republicans or Democrats in this country the day after Pearl Harbor. There were no isolationists or internationalists. And above all there were no defeatists of any stripe. There were just Americans anxious to roll up their sleeves and wade into the enemy. We are now facing a challenge fully as great as the challenge of Pearl Harbor. In some respects, it is an even greater challenge because we probably do not have as much time."[5] Johnson's willingness to race the Soviet Union contradicted Eisenhower's belief that the United States should not try to compete with the USSR in space launches. For Johnson, and contrary to what Eisenhower thought, the Soviet Union's success with the *Sputnik* satellites signaled a scientific, technological, and military advantage for the Soviet Union.

After the launch of *Sputnik,* Johnson queried the Department of Defense about the US Vanguard satellite program and the impact *Sputnik* had on the program. Defense responded to Johnson's query with a briefing on October 9, 1957, to Johnson's subcommittee. The chief of naval operations, Rear Adm. Rawson Bennett, presented the briefing with the director of Project Vanguard (John P. Hagen), and the executive director to the coordinator of the US missile program (Gen. William Bates). First Bennett, Hagen, and Bates addressed the status of the US Navy's Vanguard program and stated: "The Vanguard project has not been and is not considered part of the United States missile program. It has been handled from the beginning as a separate scientific project. It is *not* regarded as a military project."[6] Despite the relationship between satellites, nuclear warheads, and missiles pointed out by Dulles and Scoville, Bennett, Hagen, and Bates contended that the US satellite program and missile program were of little "significance" to one another. This position contradicted the public position of Eisenhower, as well as the ideas of the members of his national security staff and Lyndon Johnson. Referring to *Sputnik,* Bennett, Hagen, and Bates stated that "at the present time" they could not "give an accurate picture of the military significance of the Russian satellite."[7] The trio concluded that "none of them had given much thought to the military and political repercussions in the event the Soviets were first." Later in the briefing, after vigorous questioning from the committee's members, Bennett, Hagen, and Bates "reluctantly conceded the satellite program might have important bearing on the missile program."[8]

Despite the testimony of Bennett, Hagen, and Bates, Johnson clearly saw the similarities between military missiles and boosters for satellites, and believed that the two programs had significance to one another. Later in the decade, he advocated the development of a national space program rather than separate military and civilian space programs in an effort to reinforce the international prestige of the United States. Johnson's open identification of a national space program stood in direct contrast to Eisenhower's public position that the United States needed to maintain two separate and distinct space programs.[9]

While Johnson considered the panel's argument about missiles, satellites, and *Sputnik,* he faced a wave of correspondence from constituents in Texas. On October 15, 1957, Johnson received a letter regarding *Sputnik* from private citizen Mrs. W. A. Balendock, who stated: "The Soviet satellite indicates clearly that the Russians have caught up with the United States in terms of scientific research and technical ability. The satellite may have no military significance in and of itself, but a nation that can produce it can produce anything else."[10] In her letter, Johnson's constituent captured the sentiments of many Americans in the days and weeks after October 4, 1957. According to historian Paul Dickson, for many Americans the Soviet Union's success with *Sputnik* "created a crisis of confidence in American technology, values, politics, and the military."[11] Historian Richard Divine argues that Johnson, in an effort to respond to such sentiments, pushed for an "inquiry into why the Russians had been first" into space.[12] Johnson's efforts to investigate the failure of American technology and science to keep a lead in space served as a vehicle through which Johnson assuaged the public's concerns, while he directly challenged Eisenhower's management of the US space and defense programs. Johnson cast his investigation as a bipartisan venture and attempted to avoid "fixing blame."[13] To promote his agenda, Divine contends, Johnson worked with the White House staff and prominent Republicans such as Secretary of Defense Neil McElroy and Sen. Styles Bridges (R-NH) in an effort to "stimulate bold new thinking in defense and foreign policy."[14] In a press release in November 1957, Senator Bridges stated: "[T]he launching of Sputnik I and II should have stripped any remaining mask of complacency from the faces of thoughtful citizens. The Russian challenge is a dead earnest one in its constant quest for world domination."[15] Although key Republicans other than Bridges remained skeptical about professions of bipartisanship, Johnson conducted his inquiry professionally and was careful "not to embarrass" Eisenhower.[16] National interests and the international community's perception of America's prestige, not partisan politics, concerned Johnson.

However, some Democrats, notably Sen. Lister Hill of Alabama, recognized the political potential of Johnson's hearings and urged Johnson to use the opportunity to position himself as a leading Democratic presidential candidate in 1960. Despite Hill's advice, Johnson remained committed to reestablishing America's technological superiority.[17]

During his investigation, Johnson called prominent scientists and public figures to testify about the policies, decisions, and programs of Eisenhower and his administration. By allowing experts to directly question the path taken by the United States under Eisenhower, Johnson remained faithful to his objective of not embarrassing the president, while also astutely presenting himself as the leading critic of the Republican's space policies.

The hearings demonstrated Johnson's sympathy with the public's fears about *Sputnik* and the military capability of the Soviet Union. They also led Johnson to further toward the belief that there was a link between the military and civilian space programs of the United States. The hearings also increased Johnson's national prominence. While Eisenhower reacted to *Sputnik* by deemphasizing the significance of the event and creating NASA, Senator Johnson emerged as an advocate of a national space program. In response to Eisenhower's desire to develop NASA, the Senate established the Special Committee on Space and Astronautics to handle legislation associated with the development of the civilian space program. The committee presented a report in 1958 on Senate Resolution 3609 outlining the nation's future use of space and the conditions that would eventually lead to the banning of weapons from space. The authors of the resolution stated: "We are moved by no thought of aggrandizement. We have no intent to plant flags of conquest upon the planets or lay extensive claims to the stars. We do propose that space shall never become the route of march for tyrants and totalitarians and, as we have dedicated our resources in the past to maintain the freedom of the seas and security of the skies, so shall we dedicate our capacity to maintain the neutrality of space."[18] Later, as vice president and then president, Johnson supported the position that the United States and the international community needed to maintain space as a weapons-free frontier, and he assisted in drafting language designed to achieve this objective, but at this time the committee members spoke of the need to defend space, if necessary. In this passage, the Senate and Johnson specifically justified the civilian and military space programs of the United States. Eisenhower remained committed to the space programs as distinct and separate entities, while Johnson viewed the programs as symbiotic.

Johnson's professional and courteous handling of the hearings on *Sputnik* positioned him to become a valuable asset to Eisenhower. After Johnson's work in the Senate on space matters, Eisenhower offered him an appointment to the US delegation to a United Nations conference on the topic of banning weapons from space. Johnson accepted, believing he could "make some contribution toward making outer space an avenue of peace instead of an arena for war."[19]

Even with their partisan differences, Eisenhower and Johnson saw America's entry into space as a long-term program for solving "military and political" issues plaguing the country.[20] Both men rejected a hasty US response to the Soviet Union's successes with *Sputnik I* and *II*. In August 1958, in an article for *American Engineer*, Johnson boldly stated that the "United States must run its own course and its own race setting the pace rather than have the pace set for it."[21] This paralleled statements made by Eisenhower and his press secretary, James Hagerty, on October 9, 1957, in their respective press conferences on *Sputnik*, that "our satellite program has never been conducted as a race with other nations." Quelling public hysteria over the Soviet Union's satellite remained the paramount concern for both Eisenhower and Johnson, as neither politician wanted to admit that the United States lagged behind the Soviet Union in technological sophistication.

Eisenhower's comments in the 415th meeting of the NSC on July 30, 1959, further demonstrated that his and Johnson's thoughts on space ran parallel on some issues. In response to a series of suggestions on the path US space policy should take in comparison with the Soviet space program, Eisenhower stated that "we should not couch our own policy in terms of what the USSR is doing." Eisenhower also said that he believed the United States was not in a "one hundred yard dash, but rather a mile run" with the Soviet Union. Although Eisenhower and Johnson had the same basic ideas about the best policy path for the US space programs, they fundamentally differed on their public positions on the relationship between the military and civilian space programs. Nevertheless, the two worked together for the remainder of Eisenhower's tenure as president to improve the space programs of the United States.[22]

Johnson supported Eisenhower's pledge to keep space free of weapons, but stressed the need for the United States to recapture its international prestige. On November 12, 1958, a *Dallas Morning News* reporter quoted Johnson as saying that "peace is the first and greatest opportunity space affords." In the same article, Johnson attacked the Soviet Union's secretive space program by declaring, "In a world circled by the vehicles of space, there can be no secrets. We need, therefore, to concern ourselves not with the guarding of what we know, but rather, with exchanging and pooling our knowledge with all the community of free men."[23] Johnson's comments about the US space program and the need to work with the "community of free men" toward the sharing and development of peaceful access to space stand in stark contrast to comments attributed to him in an article in the *San Antonio News* on that same date. That article quoted Johnson as saying, "Security, of course is a concern—an urgent and imperative concern which cannot be overstated. Obviously, the whole technology of warfare will be dominated by space weapons and space vehicles."[24] Although these remarks at first seem contradictory, a deeper analysis finds common ground between them. The comment in the *San Antonio News* expressed the belief that space would become a future battleground between the United States and the Soviet Union. This reflected the persistent rhetoric of Cold War confrontation. But, in the *Dallas Morning News,* Johnson offered an alternative outcome.

By pursuing an international agreement on the banning of weapons from space, Johnson believed the United States could lead the way in dissuading the Soviet Union from weaponizing space. In fact, for the remainder of Eisenhower's presidency, Johnson and Eisenhower remained committed to the objective of working toward the space weapons ban, while striving to improve the deteriorating image of the technological and scientific capabilities of the United States.[25] Although they worked closely on space issues, a fissure emerged between Johnson and Eisenhower over domestic issues as the 1960 election approached.

The Vice President, Space, and Criticism of NASA

Having proven himself a skillful and knowledgeable leader on space issues while in the Senate, Johnson remained an active participant on space issues in the Kennedy administration. In the opening months of 1961, the new administration faced a wave of criticism from the air force over the structure and nature of the civilian and military space programs. While open to the ideas presented by the air force, Johnson believed a report outlining the administration's positions on the civilian and military space programs could provide policy guidance. Issued on June 3, 1961, the report, entitled *The National Space Program,* followed a model like that of *The Space Handbook.*[26] Unlike the latter, which was released publicly, *The National Space Program* remained a classified document. In the introduction to the report, the joint DoD-NASA team of authors reiterated a key theme in President Kennedy's May 25, 1961, message to Congress. In it, Kennedy had laid out the primary goals of the national space program and the reasons for "undertaking space projects."[27] In its space policy, the administration emphasized the scientific and commercial aspects of space, while soft-pedaling the military elements of the national space program. The administration's desire to place the emphasis on science and commercial programs is apparent in the report, the bulk of which addresses civilian space programs (ones controlled and run by NASA). High-profile civilian programs, such as the manned Mercury, Gemini, and Apollo missions, the Ranger lunar probe, and the TIROS weather satellite, exemplified the administration's commitment to scientific and commercial endeavors. These programs reinforced the administration's emphasis on the civilian elements of the national space program.

Although highlighting civilian elements, the DoD-NASA authors discussed military space programs in *The National Space Program.* Reinforcing its message, the team described these programs with language supporting the administration's public vision of space. In the section on "communication satellites," the authors highlighted the Echo II, Rebound, Westford, Relay, and Advent programs, only two of which—the USAF's Westford and the army's Advent—were military programs.[28] In the listing the "objectives" of the Westford program, the authors blandly stated that the program was designed:

- To demonstrate the feasibility of creating an artificial ionospheric layer capable of reflecting communication signal.
- To show that the orbit chaff will not materially affect other scientific pursuits in space and radio astronomy.[29]

Note that there is no mention of the military applications for communication satellites. The DoD-NASA team defined the objective of Westford in such a way as to present the satellites as peaceful and beneficial to the international community. Correspondingly, the authors defined the objective of the army's Advent program as intended "to demonstrate feasibility of a global communication system using microwave repeaters in a 24-hour Equatorial Orbiting Satellite." As with Westford, the objective specified for Advent shows no direct connection to the military benefits provided by the communication satellites.

Although Westford and Advent were military programs, the administration presented the communication satellites in *The National Space Program* with a claim that they were military research and development programs that possessed beneficial civilian spinoff capabilities for the international market. As for the other military space systems presented in *The National Space Program,* the DoD-NASA authors followed the same pattern of describing the Discoverer, Vela Hotel, Transit, and ANNA programs in a bland and general manner with an emphasis on scientific and civilian uses. As with their description of the military's communication satellites, the authors described the objectives of these satellite programs under the heading "Science and Commercial Missions," while making no mention of their possible military benefits.

According to the report, the Discoverer program provided "internal thermal environment and biomedical research."[30] Absent from the description of the Discoverer satellites was the fact that the program served as a cover for the CIA's Corona photoreconnaissance satellite.[31]

Furthermore, the report described the Vela Hotel program, a space-based detection system for nuclear explosions, as intended "[t]o obtain information and experimental data for defining and evaluating the effectiveness of an operational Far Earth Satellite Detection System which would be established in connection with US policy commitments in Geneva."[32] As they did with Discoverer, the DoD-NASA team described Vela Hotel with language emphasizing its nonmilitary applications, directly avoiding talk of the military significance and usefulness of the satellite system. As in comments on Discoverer and Vela Hotel, the authors of *The National Space Program* used the same style of descriptions to detail the navy's Transit navigation satellite and the tri-service ANNA geodetic satellite system.[33] According to the Applied Physics Laboratory at Johns Hopkins University, the Transit satellite system provided ships, submarines, and aircraft with "accurate position fixes."[34] While this predecessor to the now well-known Global Positioning System (GPS) provided valuable civilian navigation after 1967, the initial benefit of Transit was for the navy's new Polaris fleet ballistic missile submarines (SSBNs). The crews of SSBNs used Transit to get a precise fix on the submarine's position in order to calculate target data and launch points for SLBMs. Likewise, the tri-service ANNA satellite provided valuable research on the "size and shape of the earth as well as the collection of data on the earth's gravitational field."[35] Beyond the general scientific importance of researching gravitational fields, the information collected by ANNA provided data to improve the targeting accuracy of the ICBMs of the United States.

The DoD-NASA author team obscured the military applications of Transit and ANNA satellites systems to fit with the administration's emphasis on civilian space programs. Additionally, within *The National Space Program* report there is no mention of ASAT or BMD programs, even though both systems had been in various research and development phases since 1957. Using *The National Space Program* as a foundation, Vice President Johnson vigorously pursued his commitment to improve the position of the United States in the "space race" with the Soviet Union, while also advocating Kennedy's commitment to the peaceful use of space.

In order to implement the Kennedy administration's space policy, Johnson began building a consensus within the administration. Johnson petitioned Secretary of Defense Robert McNamara and NASA administrator James Webb to provide information on the national space program. On April 21, 1961, McNamara sent a memorandum to Johnson analyzing DoD's space programs. Rather than justifying the various military space programs by their explicit contribution to national security objectives, McNamara stated that "achievements in space symbolize technological power and the organizing capacity of a nation" and therefore can "contribute to national prestige."[36] With prestige in mind, as well as considering the vast economic and industrial resources necessary for a larger-scale space program to challenge the Soviet Union, McNamara accentuated the need for the space program to be "well planned and well managed."[37] Armed with his own experience and with input from McNamara, Johnson prepared a memorandum for President Kennedy entitled "Evaluation of the Space Program."

As Johnson prepared his memorandum, he also authorized the staff director and assistant director for the Senate Select Committee on Astronautics and Space Exploration, George J. Feldman and Charles S. Sheldon, to present an *Interim Report on Space Policy* to the administration. Approximately one month before Kennedy's May 1961 message to Congress, Feldman and Sheldon emphasized the need for the president to state the national goal of the US space program as intended "[t]o maintain world leadership in the peaceful exploitation of space and astronautical sciences and to insure that military applications in the area are energetically prosecuted in the interest of national security."[38] Also, emphasizing the synchronization within the emerging national space program, Feldman and Sheldon asserted: "Space programs are intimately linked with the national security. But national security is much broader than military activities, and space intricately interrelates military, political, economic, and scientific problems and needs. No one aspect can be singled out as the most important to the national security."[39] Johnson embraced the tenets advanced within the report and further established himself as an expert on space policy.

On April 28, 1961, Johnson outlined his ideas in a memo addressed to Kennedy. Johnson informed the president that the United States had failed to marshal its resources in building a space program designed to showcase "U.S. leadership" with the result that the Soviet Union had surged "ahead in world prestige attained through impressive accomplishments in space."[40] Recognizing that space could capture the attention of aligned and non-aligned nations, Johnson recommended that the United States "exploit communication, navigation, weather, and mapping [satellite programs]" as ways to counter the international image of the Soviet Union.[41] Johnson also advocated focusing a "strong effort," such as a "manned exploration of the moon," to revitalize the US space program and reestablish the attention of the international community on scientific, technological, and engineering capabilities of the United States. Aware of Johnson's recommendations, President Kennedy outlined the objectives of the national space program for members of Congress and for the nation. Although Kennedy and Johnson had successfully recast the national space program by May 1961, more work remained. On August 18, 1961, the Space Council heard presentations from Lawrence Kavanau from the Office of the Secretary of Defense (OSD), Director of Defense Research and Development Harold Brown, and Deputy NASA Administrator Hugh

Dryden on the status of the US civilian and military space programs.[42] Johnson queried the panel of Kavanau, Brown, and Dryden:

- Is the intelligence information based on agreement by the full intelligence community, or is it just based upon CIA interpretation?
- How do you visualize the Russian space threat? What form do you expect it to take and when?
- Is there a mission for men in space for defense purposes as well as for scientific or so-called peaceful purposes?
- What do you know about the Russian lunar program and how does it compare to our schedule?
- From a propaganda or prestige point of view, isn't there danger of ridicule if we put a chimpanzee into orbit? Do we need to do this? Can't we get much of the same data from instruments?
- What can you tell us of the impact of space successes on uncommitted or so-called neutral nations?
- Are we moving fast enough, and are we getting enough international advantage out of our "utility-type" space efforts? I have in mind the weather satellites, navigation satellites, and communication satellites.[43]

The queries posed to Kavanau, Brown, and Dryden demonstrated Johnson's concerns over the US position in the space race, with specific emphasis on international prestige. These overwhelming concerns advanced by Johnson remained constant in the years 1961–1967. Unlike Eisenhower, Kennedy and Johnson believed the United States had to publicly challenge the Soviet Union in an effort to improve the international image of US scientific, technological, and engineering capability.

In a sharp break from Eisenhower's policy, Johnson and Kennedy began to talk openly about the space programs of the United States as one program. As the decade unfolded, Kennedy and Johnson pushed the development of the civilian space program while becoming increasingly concerned about the Soviet Union's intentions to weaponize space. At the same time Kennedy and Johnson strove to secure a treaty banning WMDs from space, the Soviet Union in the 1960s considered using space as a medium for launching and basing nuclear weapons. In April 1962, Col. Howard Burris, USAF, sent a memorandum to Johnson concerning the "Soviet Union's interest in space warfare."[44] Burris stated that as early as 1959 the Soviet Union had begun to view the potential use of space just as "sea and air were previously utilized for delivering nuclear attacks against any place on the face of the earth." Although USAF had toyed with similar ideas and then dropped them after

1957, according to Burris, by 1962 USSR military leaders advocated stationing weapons in space.

Furthermore, Burris said that the Soviet Union linked "the doctrines of surprise attack and space warfare," to "superiority in space" which they believed would allow them to conduct a surprise nuclear attack. Ultimately Burris stressed the USSR belief that space warfare could lead to "complete victory, independent of the outcome of engagements with conventional forces on land, sea, and air."

In his analysis, Burris argued that the Soviet Union was interested in the use of space weapons and extraterrestrial warfare. After October 1957, a groundswell of public support arose in the United States for defending space from communist control. Eisenhower approved the research and development of several ASAT systems, but stopped well short of the air force proposals for space-based weapons. Eisenhower favored pursuing an international agreement on the peaceful use of space. Under the direction of Kennedy and then Johnson, the United States pursued and deployed ground-based ASAT systems while also working toward an international agreement banning weapons from space. From the American perspective, preserving space as a weapons-free frontier served as a steadfast national security objective.

Despite their interest in the use of space for peaceful purposes, Kennedy and Johnson remained constantly concerned about the Soviet interest in fractional orbital bombardment systems. Even on the eve of signing the OST, Johnson and his national security staff fretted over new intelligence about the continued interest in the development of a FOBS satellite by the Soviet Union.[45]

Compounding their concerns over Soviet bellicose intentions in space, Kennedy and Johnson dealt with a flurry of charges and accusations over the revamped space program. Sen. Barry Goldwater (R-AZ), who was a brigadier general in the Air Force Reserve, and aerospace advocate and aviation contractor Alexander de Seversky became leading critics of the administration's emphasis on the civilian space program.

Speaking before the National Rocket Club in Washington, D.C., on July 17, 1962, Senator Goldwater attacked the Kennedy-Johnson space program. Goldwater argued that America needed to "move beyond just sailing into space," because "space superiority is fundamental to the future well-being, security, and prosperity of the United States."[46] While acknowledging the scientific and technological benefits of the US space program, Goldwater questioned the Soviet propaganda about their "peaceful interests in space." Fearing that the national space program was misdirected and facilitated the

Soviet intentions to control space, Goldwater stated, "What is disturbing is our policy statements which seem to indicate, rightly or wrongly, that the bulk of our potential space efforts are wrapped up in NASA. The point I am trying to make is that the requirements of the United States for military programs in space should not be neglected in any vital aspect."[47] In addition, Goldwater cautioned, "I don't believe that it is safe to assume that the fallout from NASA space research will automatically take care of our military space requirements. And I feel further that if this is our assumption, we are gambling our national survival."[48] Although Goldwater accused Kennedy and Johnson of neglecting the military space program of the United States, they were not doing so. Yet Goldwater continued to lobby for an expanded military space program, as the air force had been advocating since 1959. Despite supporting a national space program, Kennedy and Johnson believed that America needed to emphasize its civilian program over its military program to improve the country's international image. Both Kennedy and Johnson believed that overtly bolstering the military space program of the United States contradicted the stated objectives of the administration. Goldwater disagreed, and he was not the only critic.

Two months after Goldwater's speech before the National Rocket Club, the president of the Aerospace Corporation, Ivan Getting, spoke before the same body. Getting also challenged the policies of the administration, arguing that the US military had a long history of leading expeditions, and saying he believed space should be no different. Getting asked, "Why the artificial divisions between military activities on the one hand, and all other activities for peaceful purposes on the other? Why do we associate exploration and science only with a civilian agency and deny these to the military? Why do we place an evil cast on military activities in space?"[49] Getting elaborated that the administration's "de-emphasis" of air force participation in "space experiments" damaged the "morale and mission of the military to help keep the peace." According to Getting, Kennedy might use the military as a means to "ensure the freedom of space" as previous presidents had used the US military to secure the freedom of the frontier. The administration's policies confirmed the accuracy of Getting's conclusions, but the president and vice president remained committed to using the publicly visible programs of NASA to reinforce their commitment to the peaceful use of space.

The administration faced additional criticism in fall 1962 from prominent aerospace advocate Alexander de Seversky. De Seversky wrote "Command of Space" published in *Airpower Historian* in October 1962, criticizing the state of the US space program. De Seversky underscored the importance of space, stating that "command of space will be just as decisive in any military showdown in the future as command of the air has been heretofore."[50] Extrapolating from his airpower theory, de Seversky preached that there was "no separation between air and space" and that America had a "lack of understanding of the military potential of space." As far as the administration's "slogan of space for peace," de Seversky maintained that the administration's emphasis on the use of space for peaceful purposes "only confused people" and that their continual refrain dangerously obscured the importance of space for US national security.[51]

In their attacks on the administration's emphasis on NASA over the military space program, aerospace advocates accused Kennedy and Johnson of placing the United States in a strategic disadvantage with the Soviet Union. The public rhetoric of Kennedy and Johnson created the appearance that the United States had willingly chosen to concentrate on the development of scientific and civilian programs at the expense of the military space program, as Goldwater, Getting, and de Seversky had argued. But these critics overlooked the fact that the national space program included both the military and civilian space programs. For Kennedy and Johnson the national space program worked jointly toward differing objectives. The civilian space program served as a vehicle to reestablish the nation's international prominence as the leader in science, technology, and engineering, while the military space program provided security and strategic benefits.

Even with the administration's vision, more public criticism surfaced in July 1963 when Francis Vivian Drake published "We're Running the Wrong Race with Russia!" in *Reader's Digest.* Drake criticized the path of the American space program as being too peaceable. Drake argued that the US space program stood in stark contrast to the military- and strategy-focused space program of the Soviet Union. Drake's article prompted a response from the administration. On July 22, 1963, President Kennedy wrote to McNamara and James Webb concerning Drake's article. Kennedy demanded a response to Drake's claim that "the Soviet Union is making a major effort to dominate space, while we are indifferent to this threat."[52]

Administration officials began to review the issue raised by Drake. On July 30, 1963, the director of the Arms Control and Disarmament Agency (ACDA), Charles E. Johnson, wrote to the national security advisor, McGeorge Bundy, that in his article Drake had "re-hashed" an air force argument advanced by Generals LeMay, Schriever, and Ferguson, that the United States should invest in a "large scale military space program" instead of focusing on "beating the Russians to the moon."[53] Indeed, the US Air Force worked diligently during the Kennedy administration, lobbying for a larger military space program at the expense of the moon mission, and Drake supported the air force position.

On July 31, 1963, replying to Kennedy's memorandum of July 22, Deputy Secretary of Defense Roswell L. Gilpatric claimed that Drake's article was "based for the most part on Soviet propaganda statements, faulty and greatly exaggerated interpretations of technical data, quotes by U.S. authorities taken out of context or distorted, excepts from *Air Force Magazine* articles, and the author's personal opinions and unsupported statements."[54] Gilpatric assured the president that he and McNamara were "confident that the present and projected technological base will carry forward our national space objectives while providing the flexibility we need to meet our future military contingencies as they develop."[55] The matter seemed resolved with Gilpatric's memo. However, a day later, George Reedy, special assistant to the vice president, presented Johnson with a memorandum with a completely different tone.

Reedy argued that Drake had wrongly characterized the Soviet space program as military and the US program as "peaceful." Reedy stressed that Drake seemed to have arrived at his assumption based upon the "propaganda line offered by both nations." Namely, "we have stressed that our program is for peace," while "the Soviets have stressed that their program is for military power." Reedy concluded by stating that despite the rhetoric from the United States and the Soviet Union about their respective space programs, both "programs came to approximately the same thing and the difference was entirely a matter of how one looks at it." Essentially, Reedy's interpretation was that the difference between the US and USSR space programs was only a matter of perspective.[56] Reedy's recognition of the similarity between the programs fit with the administration's desire to emphasize the civilian and scientific programs of NASA, while publicly downplaying the continued development of the military's space systems. Although Reedy's tone differed from Gilpatric's and Charles E. Johnson's, the objective to support the national space program was the same idea advocated by the administration, and specifically by Lyndon Johnson. After Johnson became president in November 1963, the criticism did not cease. In May 1964, a group of prominent Republicans led by Milton Eisenhower, the president of Johns Hopkins University, questioned the wisdom of the manned mission to the moon. The questions raised by the Critical Issues Council of the Republican Citizens Committee of the United States differed from those lodged by Goldwater, Getting, and de Seversky, as the new criticism lacked the overt partisan rhetoric about the strategic position of the United States, focusing on more practical matters.[57]

On May 28, 1964, the Critical Issues Council issued its "Critical Issues Paper No. 7," entitled *America's Space Program: An Appraisal.* In the opening paragraph, the council emphasized that the "basic objective of the United States space program should be the exploration and, ultimately, the utilization of space for peaceful purposes."[58] Despite their belief that space should be used for peaceful purposes, the council did not reject military use. In fact, the council openly endorsed the need for "an effective and economical national program" which included both civilian and military space programs, endorsing "reconnaissance, communications, meteorology, navigation, inspections, and ballistic missile defense" as the non-aggressive military uses of space. It is difficult to determine the grievances of the council from its papers alone, since they do not harshly criticize the policies and programs advocated by Kennedy and Johnson. In fact, they seemed to agree on the same basic tenets. However, the press release accompanying *America's Space Program: An Appraisal* bluntly outlined the council's disagreements with Johnson and his handling of the national space program.

The critical issues, according to the members of the council, were the timetable, cost, and benefits of the manned lunar program.[59] Concerning the "crash lunar program," the members of the council stated: "The manned lunar landing program represents a challenge, to be sure, but we do not believe that the hoped for psychological or propaganda advantages call for the present efforts to meet a publicized deadline, currently stated as prior to 1970. We find no reason to believe that putting a man on the moon could contribute to our military strength. We see no evidence that urgent defense objectives warrant a crash program or that meeting of deadlines such as 1970 serves any significant national objective."[60] The council wanted to drop the manned lunar landing deadline of 1970, and recommended that Johnson consider allowing other nations to participate in the lunar program because of the large scale and significant costs of the project.[61]

The objections and program recommendations promoted by the Critical Issues Council reached President Johnson on June 2, 1964, accompanied by a letter written by Milton Eisenhower. Johnson consulted with NASA administrator James Webb on a final draft response. After consulting his deputy Hugh Dryden, Webb recommended that the president refrain from implying that NASA had a set schedule for the moon missions, and advised Johnson to use the term "target date" instead.[62] Evidently taking Webb's advice, Johnson dictated his revised response on June 29, 1964, and thanked Milton Eisenhower for his committee's suggestions. Johnson rejected the council's wishes to disavow the deadline already associated with the manned lunar landing program and to change or reduce the structure and scale of the program.[63]

The Critical Issues Council proposals to Johnson demonstrate a shift in Republican criticism away from the hawkish tone of firebrands such as Goldwater and de Seversky. Milton Eisenhower and the council as a whole took a position supporting the development of a national space policy balancing the military and civilian space components, while also maintaining the US endorsement of the peaceful use of space.

Disarmament

By 1964, the US military firmly embraced the use of space for reconnaissance, surveillance, communication, and mapping missions. Johnson reinforced his commitment to the military use of space by recognizing that military satellites and non-aggressive space systems provided the United States with capabilities necessary to maintain and fortify national security objectives.[64] However, despite increasing Republican concern over Johnson's advocacy of the manned lunar mission, the president remained committed to using the national space program to combat the Soviet Union in the open.[65]

Johnson's endorsement of a national space program paralleled the general mood of Americans. In a poll taken in May 1964, the Johnson administration found that 48 percent of those polled supported the "equal division of spending on space between the military and non-military programs."[66] Likewise, in a second poll taken in October 1964, a majority said the same.[67] Johnson took the results of these polls as an endorsement of his position. Johnson believed it was "national policy to maintain a viable space program, not a separate program for NASA and another for Defense, and yet another for each of our several other agencies. Likewise it is understood that the United States does not have a division between peaceful and non-peaceful objectives for space, but rather has space missions to help keep peace and space missions to improve our ability to live well in peace."[68] Johnson saw a great need for the United States to ban weapons from space, while also increasing the military's use of non-aggressive satellites and space systems.[69] Like Kennedy, Johnson believed that the United States needed to keep space free of weapons, but Johnson also believed the United States needed to defend its national security by developing ASAT and BMD capabilities.

As a senator, Johnson had formed his ideas on disarmament and nonaggressive military use of space in the wake of *Sputnik*. Supporting Dwight Eisenhower's idea and honoring a request by the president, Johnson spoke at the United Nations in support of the US proposal to create a UN committee on outer space.

Wishing to stave off the contamination of space with the "distrust, fear, and ignorance" of the Cold War, Johnson believed that the United Nations could assist in maintaining space as a weapons-free frontier.[70] While key military and political leaders in the United States openly advocated the need to control and weaponize space, Johnson recognized a unique opportunity.[71]

This focus on international efforts to keep space free of weapons contradicts the traditional historiographic wisdom that the United States immediately raced the Soviet Union after the launch of *Sputnik* to place weapons in space.[72] Although Johnson and Eisenhower acknowledged that the Soviet Union had garnered prestige from its success with *Sputnik,* they realized the United States could gain political mileage by taking a public position advocating the peaceful use of space. Because of the bipolar nature of the Cold War, Eisenhower and Johnson believed that if they cast the United States as the guardians of peace, then they would automatically portray the Soviet Union as the aggressor bent on the conquest of space. It was within these parameters that Johnson first advocated keeping space free of weapons and envisioned the utility of civilian and military space programs working together to achieve this goal. During his UN speech, in reference to the need for international cooperation, Johnson stated: "If nations proceed unilaterally, then their penetrations into space become only extensions of their national policies on earth. What their policies on earth inspire—whether trust or fear—so their accomplishments in outer space will inspire also. For nations given to aggression and war and tyranny on earth, unilateral success in space technology would only multiply many times over their threat to peace. Thus it is in the interest of nations dedicated to peace and freedom that the opportunity of space not be perverted to [by] aggression and control over the earth by aggressors."[73] Johnson heralded America's commitment to peace, but also openly advocated international cooperation in space. The strategy employed by the United States after the launch of *Sputnik* was fundamentally shaped by the thinking that Johnson had brought to his presidency.

As vice president, Johnson maintained his commitment to keeping weapons from space by continually advocating the ideals of international cooperation and disarmament. In December 1963, during a speech at the Goddard Memorial Awards Dinner in Washington, D.C., Johnson stressed that "it is our policy, and will continue to be our policy, to work for joint exploration and use of space for peaceful purposes by all nations."[74] As chairman of the Space Council, Johnson emphasized the national space program as beneficial to the security and interest of the entire world. Based on this approach, Johnson believed space provided a means by which the United States and Soviet Union could reach "greater mutual understanding."

Through this mutual understanding, Johnson envisioned a greater opportunity for the United States and Soviet Union to agree on arms control regimes. Space, therefore, provided Johnson with a means he could use to advance the pacific objectives of the United States. By pursuing the need for a UN committee on space, and calling for open international cooperation in space, Johnson strove to recast the international image of the United States.

In an effort to demonstrate their resolve, Kennedy and Johnson authorized Hugh Dryden to begin negotiations with Soviet Academy of Sciences professor Anatoli Blagonravov on possible joint US-USSR space projects in 1962. As of July 1962, the United States and the Soviet Union had agreed on three joint projects: "The exchange of weather data from satellites and the eventual coordinated launching of meteorological satellites. A joint effort to map the magnetic field of the earth by means of coordinated launching of geomagnetic satellites and related ground observations. Cooperation in the experimental relay of communication via Echo Satellite."[75] Although these projects were by no means on the cutting edge of technological and scientific research, they did provide common ground for the two nations to begin building a working relationship. These agreements also reinforced the US position as a nation interested in sharing scientific research with the world community.

Although Johnson welcomed the cooperation with the Soviet Union, he remained concerned about the position of the United States in the space race. In March 1963, the executive secretary of the National Aeronautics and Space Council, Ed Welsh, outlined for Johnson the position of the United States vis-à-vis the Soviet Union in the space race. In his memorandum, Welsh stated that "the United States had an impressive record" in space during 1962.[76] Welsh emphasized the flights of Mercury astronauts Glenn, Carpenter, and Schirra, the launch of the Telstar communications satellites, the Mariner Venus probe, the TIROS weather satellite, and the test of the first stage of the Saturn I rocket as "the outstanding features" of the US space program.[77]

As for the comparative data, Welsh observed that the United States had "put four times as many payloads into earth orbit" as the Soviet Union, but that the USSR had placed "more weight into earth orbit."[78] Welsh also maintained that the "ratio of success to failures" had been comparable but that the United States had an advantage in the application of "weather, communication, and navigation satellites, while the Soviet Union had more total amount of knowledge and experience in manned space flight."[79] Judging from Welsh's tone as well as his analysis, the space race was a dead heat. Johnson considered this a vast improvement over the conditions under Eisenhower.

Even with his calls for cooperation and disarmament, and a public emphasis on the civilian space program, Johnson remained committed to the non-aggressive military use of space. Johnson believed using non-agressive military satellites and space systems contributed to the stability of the strategic environment and allowed the United States to maintain its commitment to the objective of using space peacefully, a category in which, according to Welsh, the United States led the Soviet Union.

Under Kennedy and Johnson, the spending authorized for military space systems rose from $769 million dollars in FY 1961 to an estimated $1.5 billion dollars in FY 1963.[80] While a small amount of funding went to the research and development of ASAT and BDM systems, the vast majority of funds went into the construction of satellites and heavy-lift boosters. Johnson understood that the use of satellites for weather data collection, communication, surveillance, reconnaissance, and navigation not only provided the United States with necessary strategic and tactical benefits, but also provided civilian spinoff technologies the nation could use.

Johnson ensured the military's space programs meshed with the administration's decision to transform the United States into the world's leading space power. Johnson highlighted the air force Manned Orbital Laboratory (MOL) as an example of a military space program that was non-aggressive and had the potential for civilian use. While Johnson worked to bolster the military and civilian space systems of the United States, he was presented with intelligence indicating the Soviet Union's continued interest in FOBS.[81] The USSR interest in FOBS strengthened Johnson's commitment to secure space as a weapons-free frontier.

Within days after becoming president, in a speech at the Douglas Space Center in Huntington Beach, California, Johnson stressed that the United States remained committed to the "peaceful purpose of space for the good of all mankind," while also emphasizing "that our national purpose in space is peace."[82] During his presidency, Johnson achieved his objective. From the start of his tenure as president, Johnson remained deeply concerned about the status of the Soviet space program and its military capabilities in space in comparison to those of the United States. In spring 1964, Johnson's major concern was the USSR capability to intercept satellites and place nuclear missiles in orbit. The threat of the Soviet Union's FOBS program carried over to Johnson's administration.

Although the United States had already funded and deployed ASAT Programs 505 and 437, the public was unaware of this capability. Johnson remained committed to America's ASAT program, but he also wished to extend diplomatic overtures to the Soviet Union regarding the peaceful uses of space. Despite his willingness to engage the Soviet Union in negotiations, Johnson received intelligence in May 1964 that the USSR remained interested in orbital bombardment satellites.

On May 28, 1964, in a Defense Intelligence Agency (DIA) report, the director of DIA, Gen. Joseph F. Carroll, USAF, cautioned Johnson that US intelligence expected the Soviet Union to demonstrate "spectacular military capabilities."[83] Although Carroll believed that the Soviet Union had the capability to deploy ASATs and even FOBS, he judged that the "current political environment" stifled the testing and deployment of such systems. Johnson believed that Kennedy's decision to pursue an international approach toward disarming space had generated pressure on the Soviet Union to back away from plans to base weapons in space. Although Carroll reported that US intelligence on the Soviet Union's space launches by 1964 did include discussions of ASAT and FOBS technology, Johnson believed that he could pressure the Soviet Union to abandon its interest in weaponizing space.

Supporting Carroll's analysis, the Central Intelligence Agency on May 30, 1964, issued a "Forecast of Soviet Space Spectaculars in the Balance of 1964" that offered an assessment similar to Carroll's in his May 28 DIA report. The CIA emphasized that the Soviet Union continued to pursue "the conquest of space."[84] The CIA did not comment on the Soviet Union's ability to field ASAT or FOBS, focusing rather on the Soviet Union's Zond, Vostok, and Polet programs that had captured the international community's attention. The CIA failed to mention, let alone discuss, the Soviet development or use of weather, reconnaissance, or communication satellites. Rather, the CIA stressed the "propaganda and scientific [feats]" expected by the close of 1964 and remained concerned about the ability of the Soviet Union to capture further international prestige through their space program. From the CIA's perspective, it seemed that little had changed since 1957. The concern over how the Soviet Union used its space program to boost its international image led Johnson to follow a dual approach to banning weapons in space. While working toward an international agreement to ban weapons from space and encouraging international cooperation, Johnson authorized the deployment and construction of a ground-based ASAT system designed to remove hostile satellites or FOBS systems, if the need arose.

Actions taken by Johnson in the closing months of 1964 captured the essence of his dual approach. First, in September 1964, Johnson disclosed the US ASAT program to the American public.[85] Johnson justified the program as a defensive capability designed to "intercept a satellite that might be carrying a weapon to threaten U.S. national security."[86] After Johnson's disclosure, Secretary of Defense Robert McNamara held a press conference confirming that the United States had successfully tested an ASAT system.

Johnson decided to disclose the existence of an American ASAT system for various reasons. First, the public announcement informed the Soviet Union that the United States had the technology and capability to neutralize FOBS. Second, Johnson's disclosure allowed him to fend off attacks by Republicans that he was "soft on defense."[87] Johnson also eased the concerns of Americans over the Soviet Union's designs to control space. By going public about the existence of the US ASAT system, Johnson demonstrated his strong commitment to the military space program while remaining true to his objective of establishing the United States as the world leader in working toward maintaining space as a weapons-free frontier.

Acknowledging the existence of the advanced military space capabilities of the United States also enabled Johnson to negotiate more vigorously with the Soviet Union on cooperation in space since his bargaining position would look stronger. From December 1964 through September 1966, a Johnson-authorized delegation met with a group of politicians and scientists from the Soviet Union to discuss the possibility of a joint space mission.

Building on the attempts to cooperate in a civilian space mission, Johnson continued to support discussions with the Soviet Union on the benefits of joint scientific and technological programs.[88] Johnson believed that the ongoing negotiations with the Soviet Union could lead to an arms control agreement for space. Johnson reinforced this point during a press conference on August 25, 1965. While discussing the scientific and military benefits of MOL, Johnson stated: "We intend to live up to our agreement not to orbit weapons of mass destruction and we will continue to hold out to all nations, including the Soviet Union, the hand of cooperation in the exciting year of space exploration which lies ahead for all of us."[89] The negotiations with the Soviet Union on joint civilian missions served as the foundation on which the agreement to ban weapons from space was based and allowed Johnson to proclaim that the United States moved closer to achieving its objective.

Even as Johnson and his administration strove to negotiate an international treaty on outer space, the Soviet Union, despite public statements to the contrary, worked to develop orbital nuclear weapons.[90] Although the US government knew of USSR intentions to weaponize space, the Soviet delegation to the United Nations in November 1964 officially stated that "a no-bombs-in-orbit resolution would be in the interest of the world at large."[91] In December 1964 at another meeting between the US and USSR delegations, USSR Foreign Minister Andrei Gromyko stated that the space policy of the Soviet Union "had been and continued to be a policy of peace."[92] Despite working on FOBS, the Soviet Union professed commitment to the peaceful use of space in its negotiations with the United States throughout 1965 and 1966. There was some progress toward an agreement to ban weapons from space, but Johnson's concerns over FOBS remained.

Since 1961, the United States had intelligence on the Soviet Union's development of an orbital nuclear bombardment system.[93] In March 1966, Rear Adm. Chester C. Ward (retired) wrote a commentary in the policy journal *American Security Council: Washington Report* on the contradiction between the disarmament agenda of the United States and the actions and policies of the Soviet Union. In "Space Weapons: U.S. Policy vs. Soviet Achievement," Ward maintained that the Soviet Union had been working on FOBS, despite the 1963 UN Resolution banning "the placing in orbit around the earth any objects carrying nuclear weapons or any other kind of weapons of mass destruction, installing such weapons on celestial bodies, or stationing such weapons in outer space in any manner."[94] Having cited the language of the UN resolution, Ward observed that the resolution lacked "provisions for inspection, enforcement, and penalties for violations." Ward continued that the United States had long been an advocate of banning orbital weapons and adhered to the "spirit" of the UN resolution, while the Soviet Union only abided by the "technical letter of the agreement" and still wanted to weaponize space. Ward argued that the Soviet Union interpreted the UN resolution to allow the production of space weapons, but not their deployment. The USSR position on space weapons, according to Ward, ran counter to the spirit of the resolution and the US position. Ward enhanced his position by quoting the secretary of the air force, Harold Brown, in his testimony before the House of Representatives Appropriations Subcommittee on March 12, 1964, in which Brown emphasized the peaceful uses of space.

During the hearing, Brown replied to a question about US research on orbital bombardment systems: "My recollection is that we have not more than three or four people working in one of the contract organizations studying the question. We are not doing any hardware."[95] To clarify his point, Ward cited Brown again to emphasize a lack of interest in orbital bombardment system by the United States. According to Ward, Brown had said: "There are two reasons: First, it is not a very good idea. . . . Second, there is now a U.N. resolution which we subscribe to and the Soviets have subscribed to, not to put bombs in orbit. This does not prevent people from doing the development on it, but apparently neither the Russians nor we believe it is a very important strategic weapon."[96] Contrary to Brown, Ward argued that the Soviet Union recognized the potential usefulness of orbital bombardment systems and maintained an active interest in the development and deployment of such systems.

Ward observed that the United States was faithfully adhering to the UN resolution, even though the Soviet Union continued to strive for a strategic advantage over the United States by pursuing space-based weapons. Aware of warnings from people such as Ward, Johnson focused on trying to conclude an outer space treaty in the United Nations to demonstrate the US commitment to keep space free of weapons.

To increase international pressure on the Soviet Union in light of their interest in space weapons, Johnson opened the US Navy's Transit navigation satellite system to the civilian market.[97] The declassification and open use of the Transit network gave the international community significant navigational advances. Specifically, the incorporation of space-based navigation made commercial sea traffic safer. Beyond the use of navigation satellites, Johnson also advocated the use of the Nimbus weather satellites for the open international market. Together the Transit and Nimbus satellites allowed for a greater awareness of potential hazardous weather systems, and allowed ship captains to plan and adjust routes accordingly. Satellite navigation also provided a more precise fix for ships' navigators and provided an additional level of safety in case of an emergency while at sea. These two examples demonstrate Johnson's commitment to the peaceful use of space by the international community. The Transit and Nimbus programs demonstrated the Johnson administration's interest in opening classified systems that had a greater utility for the civilian world. This was the essence of the US national space program. To further his objective and finally demonstrate that the United States was indeed the international leader in peaceful uses of space, Johnson signed the Outer Space Treaty (OST) January 1967. An eighteen-year-long period during which successive US presidents strove alone to keep weapons out of space culminated on October 10, 1967, when the Outer Space Treaty went into force. In winter 1967, Johnson sent a message to the Senate supporting the ratification of the OST. Johnson boldly stated that, in the aftermath of *Sputnik* in 1957, the concerns of America had been for the "safety of our nation and for the survival of humankind."[98] Johnson observed that the dual objectives of safety and survival had become the bedrock of the US space effort. Johnson summarized his position: "Today, outer space is free. It is unscarred by conflict. No nation holds a concession there. It must remain this way. We of the United States do not acknowledge that there are landlords of outer space who can presume to bargain with the nations of the Earth on the price of access to this domain. We must not—and we need not—corrupt this great opportunity by bringing to it the very antagonism which we may, by courage, overcome and leave behind forever if we proceed with this joint adventure into this new realm."[99]

With these powerful words, Johnson urged the Senate to move beyond the challenges of the Cold War rivalry and embrace a rare opportunity to reshape the future.

To reinforce the significance of the event, Johnson highlighted the major principles of the Outer Space Treaty:

- No nation can claim sovereignty to outer space, to the moon, or to other celestial bodies.
- All nations have a right to conduct space activities.
- No one may use outer space or celestial bodies to begin a war. The rules of the United Nations Charter apply to space.
- No country may station in space or orbit around the Earth nuclear or other weapons of mass destruction.
- No country may install such weapons on celestial bodies.
- No nation may establish military bases, installations or fortifications on a celestial body. Nor may weapons be tested or military maneuvers be conducted there. The right to visit another country's installations and space vehicles on a celestial body is guaranteed.
- Astronauts are "envoys of mankind." If an astronaut lands on another country's soil, he must be returned safely, promptly, and unconditionally. Space activities and their results are to be reported for the benefit of all.[100]

While these points emphasized Johnson's pacific and international interests, they did not hinder his advocacy for the US military space program. For Johnson, the Outer Space Treaty did not inhibit the US military programs focused on space-based reconnaissance, surveillance, weather data collection, navigation, and communication. It reinforced them as the core military missions for non-aggressive use of space.

Johnson also justified the military space program by arguing that military satellites provided security and stability, saying he believed these programs supported his position on the non-aggressive use of space. The US national space program, civilian and military, was not just for the advancement and security of the United States, but benefited the world.

On January 27, 1967, representatives from the United States and Soviet Union signed the Outer Space Treaty and proclaimed the event a monumental moment in the Cold War. Despite the optimism that day, on November 3, 1967, Secretary of Defense McNamara announced during a press conference that the Soviet Union had tested components "related to the possible development of a FOBS." McNamara stated that US intelligence indicated that the Soviet Union "could achieve an IOC [initial operational capability] of the system by 1968."[101]

Despite the administration's concerns over FOBS, NCS staffer Spurgeon Keeny drafted a statement for Sen. John Sherman Cooper (R-KY) to use in hearings held by the Senate Foreign Affairs Committee. Keeny asserted that the USSR test of FOBS components "was not a violation" of the Outer Space Treaty.[102] Specifically, Keeny pointed out the treaty forbade the orbiting of "objects carrying nuclear weapons" but not the testing of "systems capable of carrying nuclear weapons." In an effort to preserve the spirit of the treaty, Keeny advocated the United States "be careful" in its accusations about suspected treaty violations, to avoid USSR "counter-charges" that the US interest in the treaty was based on gaining "tactical and political advantages."[103] Keeny and other key national security advisers such as Walt Rostow and E. C. Welsh viewed the Outer Space Treaty as a valuable tool in America's effort to maintain the weapon-free status of space, as well as boost the international image of the United States. From their perspective, the OST was not just a public relations victory for the United States, but also a valuable tool to advance the US-led position.

Republican challengers saw the treaty and the actions of the Soviet Union as example of Johnson bargaining away America's security. Despite Republican criticism, Johnson succeeded in codifying America's commitment to preserving space as a weapons-free frontier. Furthermore, he did so while maintaining a robust military space program. By 1967, Johnson presided over the development and solidification of a national space program that emphasized the utility of civilian and military space programs in concert.

Conclusion

During Johnson's tenure, NASA and the OST overshadowed the military space programs of the United States. However, Johnson's emphasis on NASA did not mean he neglected military space systems. In 1965, as the public focused on the Gemini missions of NASA, the Pentagon and NASA considered how to best use space-based assets for the war in Vietnam.[104] In the midst of the Vietnam War, NASA worked with the air force to apply NASA space research there. Together NASA and the air force believed that satellites and space systems could identify "instantaneous cloud cover, synchronize altitude communication, and locate downed pilots."[105] The NASA-USAF team also identified research in "super-sensitive seismic sensors, lightweight power supplies, and infrared technology" as potentially useful in Vietnam. Therefore, as the public focused its attention on Gemini, early Apollo missions, and the negotiations with the Soviet Union, Johnson quietly worked to improve the military space programs begun under the Eisenhower administration. The decision to emphasize the US civilian space program, while minimizing publicity about the military space program, reflected the underlying strategy of Johnson's space policy.

Johnson believed that Eisenhower's public efforts to separate military and civilian space programs had been unnecessary. Advancing an idea first identified in the Kennedy administration, Johnson advocated a single comprehensive national space program. This approach also presented Johnson with a chance to reinforce his commitment to keeping space free of weapons, while also allowing the United States to pursue the use of space for non-aggressive military purposes. Johnson's concerted efforts to solidify America's commitment to banning weapons from space dictated that the military space programs recede out of the public eye and into the shadows, while NASA's programs enjoyed the attention of the international community. To reinforce and demonstrate the willingness and openness of the American space programs, Johnson made formerly classified weather and navigation satellites available to the international civilian market. Johnson highlighted the Transit and TIROS satellites as examples of non-aggressive military space programs with direct civilian application.

Eschewing the weaponization of space, Johnson continued to confine the US military space program to the non-aggressive missions first established in 1946. Johnson touted the benefits of the non-aggressive use of space for military purposes, even as the Soviet Union embarked on the research and development of FOBS. In order to combat the Soviet potential to control space, Johnson supported the development of ground-based ASAT and BMD systems by the United States. Johnson believed the United States needed to protect itself from FOBS.

Beyond his decision to deploy ASAT and BMD systems, Johnson completed an international disarmament agreement forbidding the basing and orbiting of weapons of mass destruction in space or on celestial bodies. The international community's support for the Outer Space Treaty solidified the US commitment to the peaceful and non-aggressive military uses of space. Lyndon B. Johnson had ensured that space remained free of weapons, for the benefit of all.

Chapter 5: Continuity and Variation, 1946–1967

From the onset of the Cold War, satellites and space programs evolved to become critical elements used to further the national security objectives of the United States. The early history of US efforts to build military satellites and militarize space has received far less attention than the history of the civilian programs of NASA. This disparity slights the fact that both the military and civilian space efforts served to reinforce the fundamental objective of US presidents between 1953 and 1967 to preserve space as a weapons-free frontier for the benefit of all. Building on the satellite feasibility studies completed by RAND engineers and USAF officers between 1947 and 1952, Presidents Eisenhower, Kennedy, and Johnson identified the value of satellites for international prestige, national security, and advanced research and development of military hardware. Before the advancements in satellite and space systems made under these leaders, key people in the RAND Corporation and USAF established the basic parameters for the non-aggressive use of military satellites, as well as the idea that space needed to be maintained as a weapons-free frontier. Between 1946 and 1952, RAND's James Lipp, Louis Ridenour, and Paul Kecskemeti worked with air force officers Bernard Schriever, C. P. Cabell, and Hoyt Vandenberg to build a solid foundation providing for the eventual incorporation of satellites and space systems into the national security structure of the United States. Meanwhile, President Truman provided no direction or support for the development of satellites by the United States during his term.

In the first decade of the Cold War, while the air force worked to build a military satellite program, Presidents Eisenhower, Kennedy, and Johnson worked to ensure that space did not become another contested frontier of the Cold War arms race. Although historians have traditionally focused on the space race between the Soviet Union and the United States in writing of the space age, they have overlooked the continuity of the space policies pursued by these three presidents. Using the ideas of engineers, social scientists, strategists from RAND and the officers of USAF between 1946 and 1952, Eisenhower, Kennedy, and Johnson figured out how to devise a national space policy that emphasized the military and civilian uses of space for the benefit of all.

This reinterpretation of the development of US space policy is significant because it demonstrates that a single idea regarding the militarization and usefulness of space was maintained from Eisenhower to Johnson. This finding contradicts the popular opinion that the United States sought to weaponize space. Furthermore, it calls into question the traditional interpretation of the space race as an action-reaction paradigm. Rather, Eisenhower, Kennedy, and Johnson saw non-aggressive military satellite development and the civilian space program as a means to shape international opinion of the scientific, technological, and military capabilities of the United States. Besides acting as ambassadors of US capability, satellites provided the United States with advanced non-aggressive military intelligence-gathering platforms that were critical in assessing the strategic nuclear balance between the United States and the Soviet Union. With the various satellite systems designed for reconnaissance, communication, weather data collection, and navigation, Eisenhower, Kennedy, and Johnson militarized space, but did not weaponize it. The important distinction between militarization and weaponization enabled Eisenhower, Kennedy, and Johnson to counter the Soviet Union's increasing international prestige after its series of space firsts. Essentially, US presidents between 1952 and 1967 cast the United States as the nation dedicated to the peaceful and non-aggressive use of space. To achieve their objective, all three worked with the General Assembly of the United Nations to maintain space as a weapons-free frontier. From the perspective of all three, both the military and civilian space programs of the United States supported this goal.

For these presidents, the military and civilian space programs became more than just passive intelligence platforms and demonstrators of the country's scientific and technological capabilities. Together these space programs merged to illustrate the presidents' intentions to use space for peaceful purposes. Johnson's signing of the Outer Space Treaty (OST) in January 1967 and the entry into force of the treaty in October 1967 signaled the culmination of the nineteen-year quest to ban weapons from space and secure space as a peaceful frontier. At the end of 1967, the United States had militarized space, banned weapons from it, and re-established itself as the world's leading space power.

Beyond the thread of continuity, each president from Eisenhower to Johnson advanced his own agenda to establish the United States as the world leader in the peaceful use of space. Therefore, the initial two decades of the space age can be characterized as one of continuity and variation.[1]

Truman and the Foundation for Space

Despite vast differences among these presidents, strains of continuity persist through the various administrations. Although Truman remained personally outside the discussions taking place on the feasibility of satellites in the late 1940s, the actions taken by the US Navy, USAF, and the RAND Corporation provided a foundation for his successors to support and develop satellites and space systems in the 1950s and 1960s.

During Truman's presidency, the air force and the RAND Corporation meticulously worked to define, research, and develop the basic parameters to use non-aggressive military satellites. Of paramount concern to Truman in this initial era of the Cold War was the need to maintain a technologically advanced military. For Truman, this meant a commitment to technological superiority in general, but in practical terms it meant airpower, not space power.

Regarding airpower development for the United States, Truman accepted the recommendations of Gen. H. H. Arnold and Thomas K. Finletter, who both argued that the United States needed to increase its technological superiority in the early years of the Cold War arms race. Although the initial feasibility studies of satellites by the navy and RAND evolved during this same period, there is nothing in the record that shows Truman to have supported the development of space power. Rather, air force leaders Bernard Schriever, Hoyt Vandenberg, and C. P. Cabell worked with members of RAND outlining future uses and capabilities of military satellites. Focusing on the use of satellites in reconnaissance, weather data collection, communication, and intelligence gathering missions, members of RAND and USAF established that satellites were something different from other military hardware.[2]

Two very important standards emerged from this critical idea. First, Louis Ridenour identified military missions for satellites. The missions identified by Ridenour in 1946 became the standard by which Eisenhower, Kennedy, and Johnson militarized space.[3] Second, James Lipp identified satellites as non-aggressive military hardware. Lipp's distinction resurfaced in the space policies of all three presidents, who remained committed to the position that the militarization of space did not contradict their pledge to use space

NASA administrator James E. Webb presents former president Harry S. Truman with a collection of rockets for the Truman Presidential Library in Independence, Missouri, November 3, 1961. NASA, courtesy of nasaimages.org

for peaceful purposes. Evolving within these two standards, non-aggressive military satellites became vital elements in US efforts to preserve world peace in an increasingly tense international security environment. Thus, military satellites did not become roving "Death Stars" fixed on incinerating the world below, but rather technologically sophisticated beacons demonstrating US commitment to the peaceful uses of space.

Beyond the development of basic standards on which future presidents built the US military and civilian space programs, the air force used this era to establish its claim as the nation's space force by identifying air and space as a continuous operational medium. Beginning in the 1950s, the air force unveiled the concept of aerospace to position itself as the nation's space force, a position it has consistently maintained into the first decade of the twenty-first century.

Some emerging features of a prospective space policy were continued and developed under later administrations. However, the differences between space-related developments in the Truman era and development in later years are just as significant. Despite the fundamental work by others concerning satellites while he was president, Truman himself remained focused on the development of nuclear and conventional military hardware. Unlike his successors, who consciously incorporated satellites and space operations into the military and national security force structure of the United States, Truman supported the development of strategic airpower and the policy of containment. Interest in space and satellites resided at lower levels in the military and academic communities. Although Eisenhower, Kennedy, and Johnson saw satellites and space programs as vital in the national security and defense strategies in the 1950s and 1960s, they remained outside Truman's consideration of the elements of national power.

Eisenhower and the Militarization of Space

President Eisenhower firmly embraced the use of satellites and space for military and civilian purposes. Specifically, Eisenhower's administration built on the ideas established by the RAND Corporation and USAF in the period under Truman, and began developing military satellite programs aimed at intelligence gathering, communication, and photoreconnaissance. Eisenhower's support of these programs led to the initial US efforts to militarize space and to the framework necessary to establish a national space policy.

As part of his program, Eisenhower lobbied the United Nations to begin an international effort to ban weapons from outer space. In Eisenhower's view, the United States needed to reestablish itself as the world leader in the peaceful use and development of space.

Although historians largely credit Eisenhower with ushering the United States into the space age, the work completed by the RAND Corporation and air force between 1946 and 1952 proved fundamental to his success. The feasibility studies and research completed during the Truman era provided Eisenhower's administration with a foundation to design, develop, and build military satellites, while also working toward an international agreement to keep space free of weapons. Eisenhower's identification of the need to pursue non-aggressive military satellites as well as his pursuit of an international arms control agreement on space was a significant break from the actions of the Truman era.

Furthermore, beginning in 1957, Eisenhower faced direct challenges from the Soviet Union as the space race became a major international dynamic within the Cold War arms race. Wishing to avoid competing with the Soviet Union, Eisenhower concentrated US resources on the development of intelligence-gathering satellites and the National Aeronautics and Space Administration. The Soviet launch of *Sputnik* on October 4, 1957, provided Eisenhower and his national security staff with concrete evidence that the Soviet Union had operational ICBMs. This alteration of the strategic nuclear balance fueled Eisenhower's support of the development of advanced intelligence-gathering and reconnaissance vehicles, such as the U-2, and, later, the Corona, MIDAS, and SAMOS military satellite systems.[4]

Beyond the capabilities of the U-2, Corona, MIDAS, and SAMOS provided the United States with alternatives to violating the air space of the Soviet Union. Although his "Open Skies" proposal had stalled, Eisenhower and his national security staff reasoned that *Sputnik* provided an international precedent on the freedom of space, especially since no nation contested the satellite's overflight. From Eisenhower's perspective, the ease and lack of international protest over the orbiting satellite presented the United States with an opportunity to exploit its position on the peaceful uses of space for the benefit of all. Seeing space assets as a vital part of the national security system of the United States, Eisenhower used satellites to gain intelligence on the Soviet Union and its allies.

Even as airpower advocates expressed their concerns about Soviet designs to control space and the need for the United States to pursue space-based weapons, Eisenhower remained focused on the peaceful use of space. Specifically, he believed that space should not become a weaponized medium in the Cold War arms race and that the United States would lead the international effort to ban weapons from space, as it had done in banning weapons from Antarctica.

Remaining committed to the non-aggressive military uses of space, Eisenhower entertained the use of ASAT and BDM systems, but constantly stressed his concerns about weaponizing space. ASAT and BMD systems remained strictly research projects under Eisenhower, as satellites had been under Truman. To further the US agenda to advance the peaceful use of space, Eisenhower assisted in the establishment of NASA to demonstrate to the world the US interest in the civilian uses of space for the benefit of all. Authorizing the civilian Discoverer, Transit, TIROS, and Courier satellite programs, Eisenhower used NASA to showcase the nation's commitment to the peaceful exploration of space. Taken together, those four satellite systems provided biomedical, navigational, meterological, and communication technologies beneficial to the international civilian community.

While advocating the benefits of NASA's civilian programs, Eisenhower steadfastly supported the non-aggressive militarization of space. In his thinking, the programs of NASA and those of the military both served to reinforce his commitment to keeping weapons out of space. Although Eisenhower oversaw the development of America's civilian and military space programs, he never wavered from his commitment to keeping space open only to peaceful uses.

Eisenhower established the structure for the United States to become, eventually, the world's leading space power. By establishing America's first space policy and NASA, Eisenhower firmly entrenched space as a permanent part of America's national security strategy. Eisenhower's commitment to the peaceful use of space also provided Presidents Kennedy and Johnson with a robust structure to use as a basis for negotiations with the Soviet Union and the United Nations to ban weapons from space. Beyond the policies of massive retaliation and the "New Look," the Eisenhower era needs to be remembered as the period in which the United States firmly committed itself to the use of military and civilian space systems in an effort to preserve space for peaceful purposes.

Kennedy: NASA and FOBS

Despite the brevity of his administration, John F. Kennedy profoundly influenced the space programs and the space policy of the United States. Against the Soviet Union's stunning success in space, Kennedy strove to recast the international image of the United States as the world's leading space power by focusing attention on the civilian space program. Kennedy highlighted the programs and objectives of NASA, but remained deeply committed to the use of space for military purposes and vigorously pursued an international agreement banning weapons from space.

Breaking with Eisenhower's policy of not overtly racing the Soviet Union, Kennedy boldly announced his idea of a manned mission to the moon. Kennedy believed his challenge would become the vehicle by which the United States would reestablish itself as the world's scientific and technological leader.

Kennedy's decision to advocate a national space program in which NASA received the bulk of attention did not mean that he allowed the US military space program to wither. Rather, Kennedy embraced the military's use of satellites and further committed the nation to the use of space. While his administration canceled the costly and troubled Dyna-Soar and SAINT programs, Kennedy supported the use of military satellites for photoreconnaissance, early warning, intelligence gathering, communication, navigation, and weather data collection. Satellite systems such as Discoverer/Corona, MIDAS, Advent, Transit, TIROS, and Nimbus became the hallmark of Kennedy's commitment to the non-aggressive military use of space.

The dual-use capability of the above-mentioned satellite systems allowed Kennedy to reinforce his commitment to the peaceful use of space. Justifying these non-aggressive military satellites as vital to the national security and seeing them as necessary to maintain international stability in the arms race of the Cold War, Kennedy pushed forward with plans to work within the United Nations to ban weapons from space. In so doing, he maintained that the non-aggressive US military satellite systems provided potential civilian benefits. Moving beyond justifying the US commitment to ban weapons from space, Kennedy began negotiations within the United Nations and with the Soviet Union, with an aim toward achieving his objective.

As the Soviet Union increasingly demonstrated intentions to weaponize space, expressing interest in the development and deployment of orbital bombardment satellites, Kennedy responded by spearheading an international drive to ban weapons from space. He also authorized the development of the ground-based 505 and 437 ASAT programs. Kennedy and his national security staff viewed ground-based ASAT weapons systems as non-aggressive and defensive in nature and this allowed them to remain committed to the message of using space for peaceful purposes, while also preparing a defense against the potential USSR threat.

Breaking with Eisenhower, Kennedy challenged the Soviet Union by advocating a manned mission to the moon. He believed that the United States could gain a significant psychological advantage in the international strategic balance by displaying the scientific and technological capabilities of the US civilian space program. Publicly, NASA became the center of Kennedy's national space program, while the military space program remained in the shadows. Under this paradigm, America's expanded civilian space program served as an international gesture to demonstrate the US commitment to the peaceful use of space. Kennedy publicly supported the civilian uses of space while quietly expanding the military use of space. He sought to convince the world of America's noble intentions, while also ensuring that the United States maintained a capability to keep space free of weapons.

Johnson and the Outer Space Treaty

President Johnson led the United States to the close of the initial phase of the space age. By the end of his presidency, the United States had successfully negotiated an international agreement banning weapons in space and possessed a limited ground-based ASAT capability to counteract any attempts by the Soviet Union to weaponize space. Under Johnson's leadership, the United States became the international leader in the use of space for peaceful purposes.

Building upon the space programs established and supported by Eisenhower and Kennedy, Johnson pushed to achieve his objective. Since the days immediately after *Sputnik,* Johnson believed that Eisenhower's support and endorsement of separate civilian and military space programs was a bad idea. Johnson supported the development of a national space program designed to balance the capabilities and objectives of the civilian and military space programs of the United States with national security objectives. Although Eisenhower also viewed military satellites and civilian space systems as vital components of the national security system, he publicly supported the division of the military and civilian efforts. Despite support for both programs, Johnson promoted NASA's program while quietly supporting the non-aggressive military use of space. Both presidents supported the same goal—to establish the United States as the world leader in the peaceful use of space.

To support his agenda, Johnson opened NASA's TIROS, Transit, and Nimbus satellites to public use for communication, navigation, and weather data collection. Johnson believed that these once-secret military satellite programs provided great civilian potential. His actions supported the official US position that space should be used for peaceful purposes.

Johnson's decision to emphasize the civilian side of the national space program did not mean that he was neglecting the military's space systems. Under Johnson, NASA and the air force worked together secretly to apply space-based assets to the war in Vietnam, while the American public fixed its attention on Gemini and Apollo. According to Johnson, the United States could simultaneously improve its military capabilities by using space-based assets while boosting the international perception of the scientific and technological capabilities of the United States through NASA's missions. Used in this way, both the military and civilian programs contributed to the strength of the United States in its struggle with the Soviet Union. By using the national space program as a medium in which to demonstrate US scientific, technological, and engineering capabilities, Johnson succeeded in reversing the less-than-stellar international image of the United States.

Johnson saw space exploration as an opportunity for the United States and Soviet Union to move beyond the arms race that characterized the Cold War, to arrive at an agreement for the benefit of the whole world. In this context, space had to remain free of weapons.

While Johnson worked with the United Nations to secure an international agreement banning weapons in space, the Soviet Union continued researching FOBS.[5] Informed by intelligence about the continued USSR interest in FOBS and concerned about the Soviet Union's potential to deploy nuclear-armed satellites, Johnson advocated the development of ground-based ASAT and BMD systems, believing that they did not violate his pledge not to use space weapons. For him the simultaneous support of ASAT and BMD systems and an international treaty to ban weapons from space was no contradiction. He believed they both supported his goal of using space for peaceful purposes.

Johnson's signature on the OST in January 1967 signaled the end of a two-decade quest by the United States to keep space free of weapons. At the same time, non-aggressive military satellites became valuable assets in the US security and defense strategy. Under the leadership of Johnson, the United States expanded its use of non-aggressive military satellites and secured space as a weapons-free frontier.

Acknowledging Johnson's support of a national space program and his work on completing the OST shatters the myth that the United States was interested in "arming the heavens" during his presidency. As noted above, he, like Eisenhower and Kennedy, never sought to weaponize space, but worked to keep it from becoming the next contested battleground in the Cold War arms race.

Conclusion

The foundation established in the first phase of the space age provided the parameters within which the US national space program flourished. Presidents from Richard Nixon to Ronald Reagan, like their predecessors, adhered to the belief that space should be used for peaceful purposes and worked to expand the non-aggressive use of military space systems.

This second period of the space age is characterized by its continuation of principles established between 1946 and 1967. However, the basic ideas that remained current for roughly thirty years eroded after March 23, 1983. Starting with President Reagan's televised SDI speech on that date, policy makers, strategists, and defense experts began to rethink the use of weapons in space. Although the current debate on the weaponization is beyond the scope of this book, the principles and arguments advanced by both sides parallel many of the arguments and ideas encountered by Eisenhower, Kennedy, and Johnson.[6] As the new debate on the weaponization of space rages, the parties on both sides lack a detailed knowledge of the history of that guided previous administrations in their development of space policy.

The history of the development of US space policy between 1946 and 1967 provides critical insights to the current debate. Beyond its potential to provide information relevant today, it has additional significance. First, the traditional characterization of the space age as a phenomenon of the Eisenhower years needs reconsideration. Although the military and civilian space programs blossomed under Eisenhower's presidency, employees of RAND and officers in the air force during the Truman administration defined basic use of satellites for non-aggressive military and civilian missions. The work done by RAND's Louis Ridenour, James Lipp, Paul Kecskemeti, S. M. Greenfield, and W. W. Kellogg outlined the military and civilian uses of satellites between 1946 and 1951. Working closely with RAND, USAF officers H. A. Craig, Hoyt Vandenberg, Charles Cabell, and Bernard Schriever defined the specifications and technical parameters necessary for the development of military satellites during the same period. This foundation proved pivotal in Eisenhower's decision to militarize space. Therefore, to consider that the space age began during the Eisenhower presidency is to ignore the conceptual work completed during Truman's administration. The parameters that historians have used to define the space age need to be expanded to include the work done in the years between the end of World War II and the start of Eisenhower's presidency.

The second significant element of the history of the development of US space policy between 1946 and 1967 is that the United States did not seek to occupy and claim space as its own strategic frontier. Rather, Eisenhower, Kennedy, and Johnson worked to ensure that space remained a scientific frontier open to all nations. Although all three presidents faced criticism over their decisions to not base weapons in space, they firmly believed that it was in the best interest of the United States and the world to work toward an international agreement securing space as a weapons-free frontier.

The development of the policy to use space for peaceful purposes, which included the use of non-aggressive military satellites, challenges the historiographic interpretation that the United States simply reacted to the actions and policies of the Soviet Union during the Cold War. Instead, Eisenhower, Kennedy, and Johnson developed space policy built on a consistent idea that space served to benefit all nations.

All three presidents believed it was crucial to work through the United Nations to demonstrate that the United States led the world in the peaceful use of space for military and civilian purposes, and they believed in the need for an international treaty to ban the stationing of weapons in space. Therefore the arms race paradigm associated with the strategic nuclear balance of the Cold War does not apply to the space systems and policies developed by the United States between 1952 and 1967. It is necessary to acknowledge that the United States did seek to gain a strategic advantage over the Soviet Union through its space program. However, the strategic advantage sought by Eisenhower, Kennedy, and Johnson was not predicated upon building bigger and more numerous space systems. Rather, the presidents strove to establish the United States as the world leader in the peaceful uses of space by advocating a program consistent with their stated objectives. In the context of this analysis, national space policy became a critical element in US national security strategy.

The final significant element in the development of US space policy in the first era of the space age was the dynamic interaction between the civilian and military space programs. Although the air force tried to establish itself early on as the nation's space force, by 1957 NASA had obtained functionally equal autonomy and influence, even though the air force had spearheaded the initial research and development for satellite and space systems.

In contrast to the Soviet Union's military-led space program, Eisenhower believed that the United States could use NASA to cast a different international image on the use of space. Although the military and civilian sides of the US space program saw themselves as being in direct competition, Eisenhower privately (and Kennedy and Johnson publicly) viewed the two programs as working in concert. Despite their rocky and at times adversarial relationship between 1958 and 1962, the two programs were brought in line by President Kennedy with his notion of unified national space program. Although this idea differed from the public notion of two independent programs that Eisenhower had supported, the unified national space program became the standard.

The rivalry and turf battles that characterized the relationship between USAF and NASA faded in the 1960s as both elements of the national space program worked to establish the United States as the world leader in the peaceful use of space. By October 1967, the United States had reached this goal and had negotiated an international treaty that banned weapons from space.

Between 1946 and 1967, the United States did not seek to race the Soviet Union to arm the heavens, but rather strove to develop a military and civilian space program and policy that advocated the use of space for peaceful purposes. Presidents Eisenhower, Kennedy, and Johnson used this objective to develop and deploy both military and civilian satellites and space systems. At the end of the first era of the space age, the United States had succeeded in developing a space policy that affirmed the pledges made by these presidents to preserve space for pacific uses. Ultimately the evolution of the US space policy between 1946 and 1967, the first era of the space age, can be characterized as an ongoing commitment to the peaceful use of space for the benefit of all.

Appendix A: Chronology of Significant Events, 1945–1967

10/03/45	US Navy first assesses feasibility of a satellite with North American Aviation and the Guggenheim Aeronautical Laboratory
05/02/46	RAND Division of Douglas Aircraft Company completes feasibility study of a satellite for the Air Force
05/13/46	General Eaker sends memorandum titled "A Projectile to the Moon" to Assistant Secretary of Defense Symington
07/26/47	National Security Act passes
01/12/48	Lt. Gen. H. A. Craig, USAF, declares a satellite will soon become fiscally feasible
01/15/48	Gen. H. S. Vandenberg issues policy statement on USAF interest in space
01/28/49	RAND holds internal conference to discuss the use of psychological weapons; discussion includes potential use of satellites
10/04/50	RAND publishes Paul Kecskemeti's "The Satellite and Rocket Vehicle: Political and Psychological Problems"
03/17/51	Gen. C. P. Cabell, USAF, outlines technical requirements for a USAF intelligence-gathering satellite
04/01/51	S. M. Greenfield and W. W. Kellogg of RAND publish "Inquiry into the Feasibility of Weather Reconnaissance from a Satellite Vehicle"
10/30/53	NSC 162/2 becomes basic national security policy for the United States
03/27/54	Technology Capabilities Panel (TCP) formed to investigate use of technology to avert surprise attack from USSR
02/14/55	TCP publishes its report
05/20/55	NSC 5520 outlines the objectives of the US scientific satellite program

05/27/55	Eisenhower endorses the satellite program outlined in NSC 5520
10/03/56	NSC issues a progress report to Eisenhower on NSC 5520
11/09/56	CIA advisory committee informs Eisenhower that USSR could launch a satellite before the United States
01/24/57	During 310th meeting of the NSC, Eisenhower receives further intelligence on advanced state of USSR satellite program
10/04/57	Soviet Union launches *Sputnik*
11/03/57	Soviet Union launches *Sputnik II* with a dog on board
11/08/57	US Army directed to use Jupiter C to launch Navy's Vanguard satellite
11/25/57	Preparedness Investigating Subcommittee of Senate Committee on Armed Services begins investigating US satellite and missile program
11/29/57	Gen. Thomas White, USAF chief of staff, outlines aerospace concept
12/06/57	First unsuccessful attempt by the United States to launch Vanguard
01/01/58	Alexander de Seversky article written for USAF proclaims "air power is space power"
01/16/58	John Foster Dulles, speaking at United Nations, declares United States is interested in use of space for peaceful purposes, not war
01/28/58	Gen. Homer Boushey declares space is a natural extension of USAF air operations
02/01/58	United States launches *Explorer I*
02/05/58	Second unsuccessful attempt to launch Vanguard
02/07/58	Advanced Research Projects Agency (ARPA) created in DoD
03/17/58	United States successfully launches *Vanguard I*
04/02/58	Eisenhower proposes establishment of NASA
05/15/58	Soviet Union launches *Sputnik III*, the first orbiting geophysical laboratory
07/29/58	Eisenhower approves National Aeronautics and Space Act of 1958, establishing NASA
08/18/58	NSC 5814/1 establishes US policy on outer space
10/01/58	NASA activated
11/12/58	Sen. Lyndon Johnson in *Dallas Morning News*: "peace is the first and greatest opportunity space affords us"; in *San*

	Antonio News: "[the] whole technology of warfare will be dominated by space"
11/17/58	Senator Johnson, at request of President Eisenhower, speaks before UN General Assembly to urge creation of ad hoc space committee
12/17/58	Project Mercury publicly announced
12/18/58	United States launches SCORE, world's first active communication satellite
02/28/59	First attempt to launch Discoverer satellite
04/13/59	First successful launch of Discoverer satellite
08/07/59	*Explorer VI* provides first television picture from space
09/12/59	Soviet Union achieves first lunar impact: *Luna II*
10/04/59	Soviet Union launches *Luna III* which provides first photos of moon's far side
02/26/60	United States launches *Midas I*
04/01/60	United States launches *TIROS I*, world's first weather satellite
04/13/60	United States launches *Transit IB*, world's first navigation satellite
05/01/60	U-2 and pilot Francis Gary Powers shot down by Soviet Union
05/24/60	United States launches *Midas II*
08/10/60	First successful payload recovery from *Discoverer* satellite
08/12/60	United States launches *Echo I*, first passive communication satellite
10/04/60	United States launches *Courier IB*
10/11/60	United States launches *Samos I*
11/23/60	United States launches *TIROS II*
01/31/60	United States launches *Samos II*
01/30/61	President Kennedy's State of the Union address
02/27/61	Last launch of a Discoverer satellite by the United States
04/21/61	Yuri Gagarin becomes first man in space, aboard *Vostok I*
04/24/61	George J. Feldman and Charles S. Sheldon publish *Interim Report on Space Policy*, prepared on Lyndon Johnson's request
05/05/61	Alan Shepard becomes first American in space, aboard Mercury's *Freedom 7*
05/25/61	President Kennedy outlines his space program in special address to Congress

06/03/61	Kennedy administration publishes *The National Space Program*
06/19/61	USAF issues AFOS 2/2 "Long Range Concepts as to the Nature of Future Warfare, USAF's Views on Military Activity in Space"
07/12/61	United States launches *Midas III* and *TIROS III*
07/21/61	Virgil Grissom launched into space aboard Mercury's *Liberty Bell 7*
09/09/61	United States launches *Samos III*
09/20/61	USAF publishes *The Space Plan*
10/18/61	Chairman of Joint Chiefs of Staff, Gen. Lyman L. Lemnitzer, US Army, advocates need for "defense against missiles"
12/04/61	USAF establishes task force for "follow-on" to *The Space Plan*
01/03/62	NASA announces Project Gemini
02/08/62	United States launches *TIROS IV*
02/12/62	Gen. James Ferguson, USAF, briefs Congress on the "follow- on" to *The Space Plan*
02/20/62	John Glenn orbits Earth aboard Mercury's *Friendship 7*
05/24/62	M. Scott Carpenter enters space aboard Mercury's *Aurora 7*
06/12/62	United States launches *TIROS V*
07/09/62	Starfish Prime nuclear test series provides data on EMP
07/17/62	Sen. Barry Goldwater publicly criticizes Kennedy space program in lecture before National Rocket Club
08/12/62	*Vostok I* and *Vostok III* achieve first rendezvous in space
09/12/62	President Kennedy announces manned lunar mission in speech at Rice University
09/12/62	Gen. Bernard Schriever, USAF, proposes ASAT program to Secretary of the Air Force Eugene Zuckert
09/15/62	ACDA provides President Kennedy with memorandum outlining possible arms control measures for outer space
09/18/62	United States launches *TIROS VI*
09/18/62	Ivan Getting attacks Kennedy's space program in speech before the National Rocket Club
09/20/62	Under Secretary of the Air Force Joseph V. Charyk announces the re-designation of Dyna-Soar, now the X-20

09/21/62	Under Secretary of State U. Alexis Johnson circulates memorandum to UN's General Assembly outlining "general guidance" on space disarmament
10/03/62	Walter Schirra launched into space aboard Mercury's *Sigma 7*, orbits for nine hours
10/31/62	United States launches *ANNA IB*, the first geodetic satellite
11/20/62	Secretary of Defense Robert McNamara authorizes USAF proposal for a national ASAT system
12/13/62	United States launches the communications satellite *Relay I*
03/26/63	The executive secretary of NASC, Ed Welsh, outlines for Lyndon Johnson the US position in the space race
05/09/63	United States launches *Midas VI*, as part of BMEWS
05/15/63	L. Gordon Cooper launched into space aboard Mercury's *Faith 7*, spends twenty-four hours in orbit; the last Mercury mission
05/24/63	US Ambassador to Czechoslovakia Outerbridge Horesy officially objects to a FOBS display presented by the Soviet Union
06/19/63	United States launches *TIROS VII*
07/22/63	President Kennedy expresses concern over article in *Reader's Digest* entitled "We're Running the Wrong Race"
07/27/63	First US ASAT systems, Program 505, obtain IOC
10/17/63	United States launches first nuclear detection satellite (part of Vela Hotel series)
12/10/63	Secretary of Defense Robert McNamara publicly announces the cancelation of the Dyna-Soar/X-20 program and startup of USAF MOL program
12/21/63	United States launches *TIROS VIII*
01/21/64	United States launches *Relay II*
05/28/64	CIC issues and sends President Johnson policy paper titled *America's Space Program: An Appraisal*
05/28/64	DIA reports to Johnson that United States could expect to see "spectacular military activity" from the Soviet Union
05/30/64	CIA issues "Forecast of Soviet Space Spectaculars in the Balance of 1964"
06/17/64	United States launches *Vela Hotel III* and *IV*

06/29/64	President Johnson writes a personal response to CIC and Milton Eisenhower
08/28/64	United States launches *Nimbus I*
10/12/64	*Voskhod I* becomes first orbital mission with more than one crew member
11/20/64	Foreign Minister Andrei Gromyko states that space policy of USSR has been and continues to be a policy of peace
01/22/65	United States launches *TIROS IX*
03/18/65	Cosmonauts from *Voskhod II* conduct first space walk
03/23/65	United States launches *Gemini III*, the first manned *Gemini* mission
06/02/65	United States launches *TIROS X*
06/03/65	United Stated launches *Gemini IV*
06/20/65	United States launches *Vela Hotel V* and *VI*
07/16/65	Soviet Union launches *Proton I*, the first cosmic ray station
08/21/65	United States launches *Gemini V*
11/16/65	Soviet Union launches *Venera III* which achieves first impact on Venus
12/04/65	United States launches *Gemini VII*
12/15/65	United States launches GEMINI VI-A
01/31/66	Soviet Union launches *Luna IX* which provides first pictures from lunar surface
03/16/66	United States launches *Gemini VIII*
03/31/66	Soviet *Luna X* becomes the first lunar orbiter
05/15/66	United States launches *Nimbus II*
06/03/66	United States launches *Gemini IX-A*
07/18/66	United States launches *Gemini X*
09/12/66	United States launches *Gemini XI*
11/01/66	United States launches *Gemini XII*, last of the manned *Gemini* missions
12/21/66	Soviet *Luna XIII* conducts first tests of the lunar surface
01/27/67	President Johnson signs Outer Space Treaty
04/28/67	United States launches *Vela Hotel VII* and *VIII*
10/10/67	Outer Space Treaty enters into force
11/03/67	Secretary of Defense Robert McNamara announces that United States had intelligence indicating USSR has tested components of a FOBS

Appendix B: US Space Spending

Figure B-1

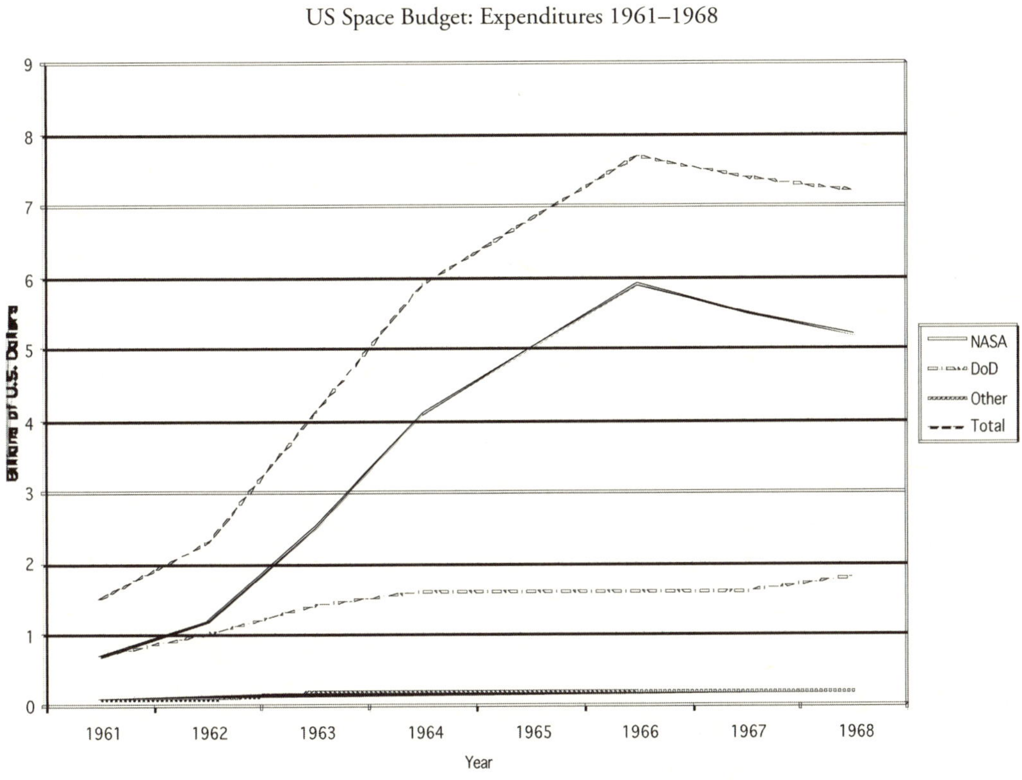

Source: Compiled from data found in National Aeronautics and Space Council, *Report to the Congress from the President of the United States: United States Aeronautics and Space Activities 1965* (Washington D.C.: Government Printing Office, 1966), 166–67; National Aeronautics and Space Council, *Report to the Congress from the President of the United States: United States Aeronautics and Space Activity 1966* (Washington D.C.: Government Printing Office, 1967), 166–67.

Figure B-2

US Space Activities: Historical Budget Summary, FY 1955–1968

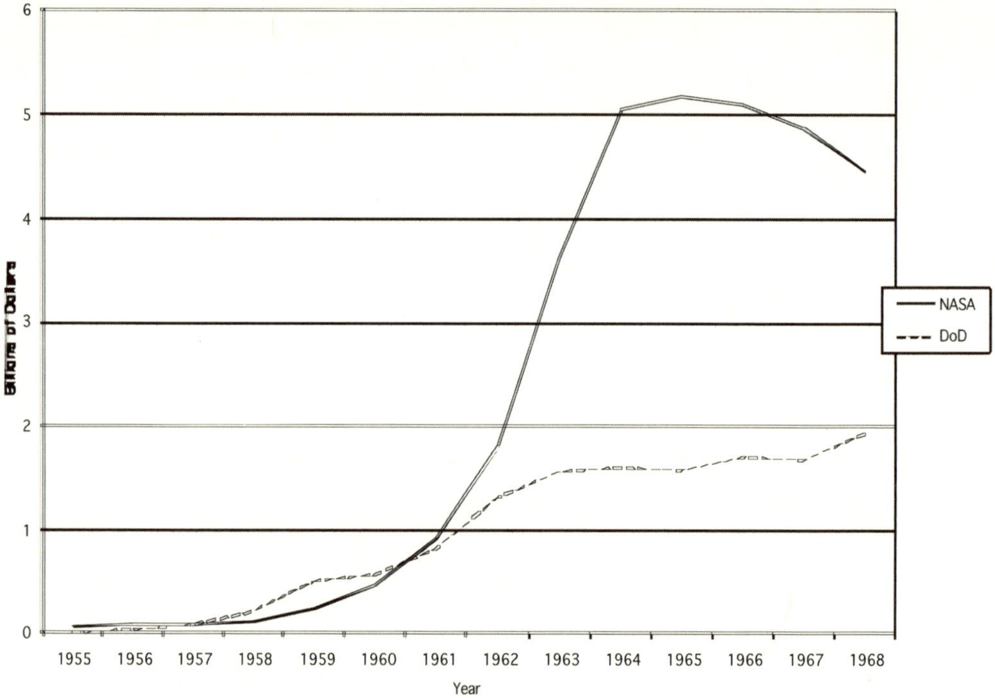

Source: Complied from data found in Spires, *Beyond Horizons*, 294; NASC, *United States Aeronautics and Space Activities, 1965*, 166–67; NASC, *United States Aeronautics and Space Activities, 1966*, 166–67.

Figure B-3

Percent Total Amount Spent on Space Activities by the US Government, FY 1959–1968

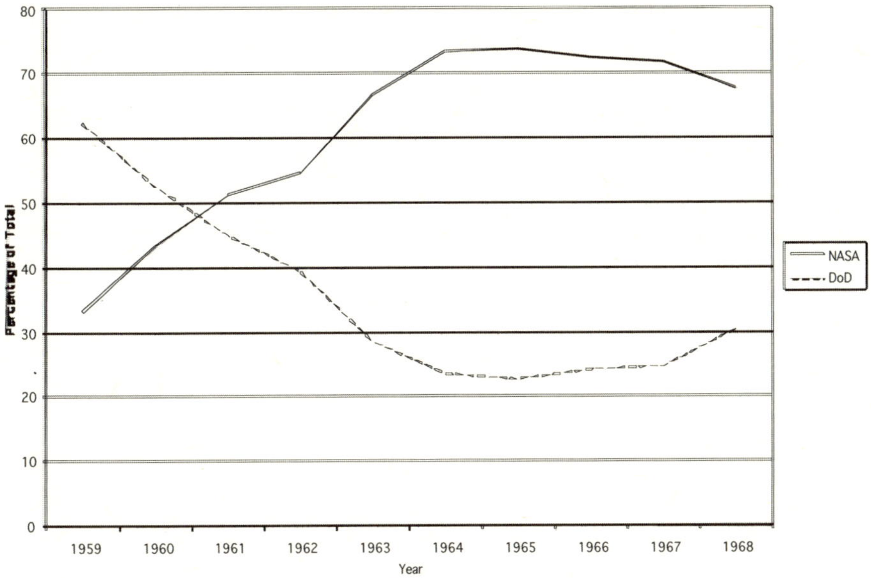

Source: Complied from data found in Stares, *The Militarization of Space*, 225; Spires, *Beyond Horizons*, 291.

Figure B-4

US Space Budget: New Obligation Authority 1961–1968

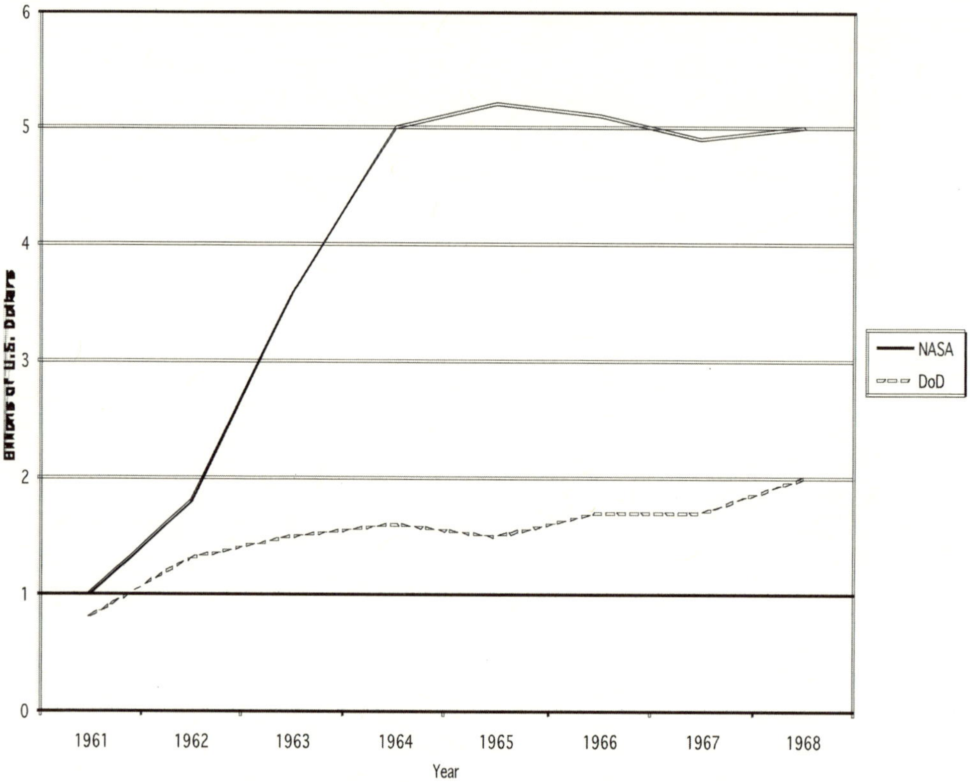

Source: Complied from data in Spires, *Beyond Horizons*, 295; NASC, *United States Aeronautics and Space Activities, 1965*, 1966; NASC, *United States Aeronautics and Space Activities, 1966*

Notes

Introduction

1. Chris Vaughn, "General: Weapons Not Goal," *Star-Telegram,* July 28, 2005, 1.

2. A case can be made for the occasional usage of "weapon-free"; however, to remain consistent throughout the writing, I will use "weapons-free."

3. Roger D. Launius, "Preface and Acknowledgment," in *Reconsidering Sputnik: Forty Years since the Soviet Satellite,* Roger D. Lauinus, John M. Logsdon, and Robert W. Smith, eds. (London: Routledge, 2002), ix–xiv; Rip Bulkeley, *The Sputnik Crisis and Early United States Space Policy: A Critique of the Historiography of Space,* (Bloomington: Indiana University Press, 1991), 3; Matt Bille and Erika Lishock, *The First Space Race: Launching the World's First Satellite* (College Station: Texas A&M University Press, 2004), 3.

4. The best work on the civilian space programs of the United States is the multivolume *Exploring the Unknown: Selected Documents in the History of the U.S. Civil Space Program,* edited by John M. Logsdon, Dwayne A. Day, and Roger Launius, (Washington D.C.: NASA History Office, 1996); The single most comprehensive coverage of the space programs of the Soviet Union can be found in Asif A. Siddiqi, *Challenge to Apollo: The Soviet Union and the Space Race, 1945–1974* (Washington D.C.: NASA History Office [hereafter NASAHO], 2000).

5. Paul B. Stares, *The Militarization of Space, U.S. Policy 1945–1984* (Ithaca: Cornell University Press, 1985); Norman Friedman, *Seapower and Space: From the Dawn of the Missile Age to Net-Centric Warfare* (Annapolis: Naval Institute Press, 2000); Donald R. Baucom, *The Origins of SDI, 1944–1983* (Lawrence: University Press of Kansas, 1992); Jeffrey T. Richelson, *America's Space Sentinels: DSP Satellites and National Security* (Lawrence: University Press of Kansas, 1999); David N. Spires, *Beyond Horizons: A Half Century of Air Force Space Leadership* (Maxwell Air Force Base: Air University Press, 1998).

6. Launius and McCurdy, in *Spaceflight and the Myth of Presidential Leadership,* Urbana: University of Illinois Press, 1977, argue since Eisenhower presidents have consistently relied on "presidential edicts" to advance the agenda of the US space programs. Borrowing heavily from the political scientist Richard E. Neustadt, Launius and McCurdy focus on policy. While useful in understanding the history of the US space programs, this model falls short because it fails to consider the role of the military space program and the US agenda to ban weapons from space. The theory of continuity advanced by Launius and McCurdy stands in direct contrast to the theory of Linda Krug, who, in *Presidential Perspectives on Space Exploration,* argues that the history of the US space program has been characterized by "disarray" and lack of continuity.

Chapter 1

1. Stephen M. Rothstein, "Dead on Arrival? The Development of the Aerospace Concepts, 1944–1958," (master's thesis, School of Advanced Airpower Studies, November 2000), 21–27.

2. The history of missile and rocket booster development coincided with the development of satellites, but it is beyond the focus of this book. For a history of the US ballistic missile program, see G. Harry Stine, *ICBM: The Making of the Weapon that Changed the World* (New York: Orion Books, 1991); Kenneth F. Gantz, ed., *The United States Air Force Report on the Ballistic Missile* (New York: Doubleday, 1958); Neil Sheehan, *A Fiery Peace in a Cold War: Bernard Schriever and the Ultimate Weapon* (New York: Random House, 2009).

3. Robert Perry, *Origins of the USAF Space Program, 1945–1956* (United States Air Force, Space Systems Division, 1961); George Washington University, National Security Archive (hereafter GWUNSA), Military Uses of Space Collection, 1945–1992, Document # 00313, 9–10.

4. Paul Stares, *The Militarization of Space* (Ithaca: Cornell University Press, 1985), 25.

5. Frank Collbohm, interviewed by Martin Collins and Joseph Tatarewicz, July 28, 1987, transcript, Joint Oral History Project of the RAND Corporation, National Air and Space Museum [NSAM], Washington D.C., 19; Bruno Augenstien, interviewed by Martin Collins and Joseph Tatarewicz, July 28, 1986, transcript, Joint Oral History Project of the RAND Corporation, NSAM, Washington D.C., 7; Stares, *The Militarization of Space,* 25. In contrast to the other sources, Stares indicates that the Douglas Aircraft Company also was included in the initial rounds of the US feasibility studies. However, this is the only source that lists the Douglas Company's involvement in the US Navy's satellite interest.

6. Stares, *Militarization of Space,* 25.

7. Perry, *Origins of USAF Space,* 10.

8. Bernard A. Schriever, "Military Space Activities: Recollections and Observations," in *The U.S. Air Force in Space: 1945 to the Twenty-first Century,* R. Cargill Hall and Jacob Neufeld, eds. (Washington D.C.: United States Air Force, 1998), 13.

9. Rothstein, "Dead on Arrival?," 2.

10. RAND Corporation, *The Preliminary Design of an Experimental World-Circling Spaceship* (Santa Monica: RAND Corporation, 1946).

11. Ibid., 10–11.

12. Peter Lang Hays, "Struggling towards Space Doctrine: U.S. Military Space Plans, Programs, and Perspectives during the Cold War" (PhD dissertation, Fletcher School of Law and Diplomacy, Tufts University, 1994), 61–62; Paul H. Nitze with Steven L. Rearden and Ann M. Smith. *From Hiroshima to Glasnost: At the Center of Decision* (New York: Grove Weidenfeld, 1989), 104–5.

13. Douglas Aircraft Company, *Preliminary Design,* 1–2.

14. The Douglas Aircraft Company engineers involved in the production of Report No. SM-11827 were D. Griggs, L. Ridenour, F. Clauser, G. Peebles, P. Lagerstrom, W. Klemperer, J. Lipp, R. Krueger, E. Graham, R. Shevell, V. Sturdevant, G. Grimminger, H. Luskin, B. Baker, E. Bradshaw, E. Wheaton, H. Liepmann, and D. Wall.

15. Sean N. Kalic, "Withering Triad: An Examination of the Modernization Programs for the Strategic Nuclear Forces of the United States" (master's thesis, Southwest Missouri State University, 1997), 8–19.

16. Rothstein, "Dead on Arrival?," 8.

17. Matthew Mowthorpe. *The Militarization and Weaponization of Space* (New York: Lexington Books, 2004), 13.

18. D. V. Gallery to John A. McCone, October 22, 1947, Harry S. Truman Presidential Library [hereafter HSTPL], Record Group 220, Box 41, File: MH3–4 Guided Missiles, 2.

19. Air Policy Commission to Colonel Boatner and Captain Pihl, memorandum No. 39, undated, HSTPL, Record Group 220, Box 41, File: MH3–4 Guided Missiles, 2.

20. E. C. Sweeney, "Naval Air Research and Development: Notes of Meeting with Bu-

reau of Aeronautics, September 25, 1947," HSTPL, Record Group 220, Box 40, File: Weihmiller, H. E., MG1–14 Classified Material, 1.

21. L. Parker Temple, III, *Shades of Gray: National Security and the Evolution of Space Reconnaissance* (Reston, Va.: American Institute of Aeronautics and Astronautics, 2005), 23–43.

22. Air Policy Commission Report (draft), undated, HSTPL, Record Group 220, Box 41, File: ML1–10 Proposed Classified Report, 10.

23. "Notes on Appearance of Donald M. Nelson, before the Executive Session," October 23, 1947, HSTPL, Record Group 220, Box 40, File: Weihmiller, H.E., MG1–14 Classified Material, 2.

24. Nitze, *From Hiroshima to Glasnost,* 96.

25. John A. McCone to Paul E. Pihl, October 1, 1947, HSTPL, Record Group 220, President's Air Policy Commission, Box 38, File: MC2–18 Central Intelligence Group 9/30. The basis of this letter is actually a series of questions concerning Russia's aircraft production, performance, and air force structure.

26. Gillispie Craigie to Thomas K. Finletter, October 21, 1947, HSTPL, Record Group 220, President's Air Policy Commission, July 1947–January 1948, Classified Material MA-3 to MC-4, Box 38, File: MC3–29 Air Force, U.S. (Craigie, Gillispie) 10/27, 1.

27. National Advisory Committee for Aeronautics, "National Program of Transonic and Supersonic Wind Tunnels: Presentation to the Air Policy Committee, November 19, 1947," HSTPL, Record Group 220, President's Air Policy Commission, July 1947–Janaury 1948, Classified Material MC4–11 to MF4–1, Box 39, File: MC4–11 National Advisory Committee for Aeronautics 11/19, 1; Dik Daso, *Hap Arnold and the Evolution of American Airpower* (Washington D.C.: Smithsonian Institution Press, 2000), 81.

28. Air Engineering Development Center (AEDC), "Presentation to the President's Air Policy Commission, October 27, 1947," HSTPL, Record Group 222, President's Air Policy Commission, July 1947–January 1948, Classified Material MC4–11 to MF4–1, Box 39, File: MF4–1 Information Regarding Russian Activities, 1.

29. Bradley Dewey to Dr. Vannevar Bush, April 29, 1946, HSTPL, Papers of Lawrence R. Hafstad, Subject File G-U, File: Guided Missiles- Joint Research and Development Board- Personal File, 1.

30. Lawrence R. Hafstad, "Introduction to Guided Missile Problems, December 1946," HSTPL, Papers of Lawrence R. Hafstad, Subject File G-U, File: Guided Missiles-Joint Research and Development Board- Personal File, 5.

31. The most famous dismissal of missiles as military weapons came from the director of the Office of Scientific Research and Development and chairman of the Joint Committee on New Weapons of the Joint Chiefs of Staff, Dr. Vannevar Bush. Testifying before the Senate Committee on Atomic Energy in December 1945, Bush stated, "There has been a great deal said about a 3,000 mile high angle rocket. In my opinion such a thing is impossible and will be impossible for many years.'' This quotation by Bush, led Theodore von Karman to characterize Bush as "a good man . . . limited in vision." See David Spires, *Beyond Horizons: A Half Century of Air Force Space Leadership* (Maxwell Air Force Base: Air University Press, 1998), 11–12.

32. Department of the Air Force, "Presentation to the Joint Chiefs of Staff Committee on the Actual Status of the Air Force, 1 July 1948 and the Planned Status of the Air Force 1, July 1949," HSTPL, Personal Papers of George J. Richards, "Personal Office File" 1942-FY 1951, Air Force Presentations FY 1949–1950, Box 3, File: Department of the Air Force, 6.

33. "Functions of the Armed Forces and the Joint Chiefs of Staff," HSTPL, Papers of Harry S. Truman, Official File, 1285 (April 1949–September 1950), Box 1594, File: 1285 (Apr–July 1949), 12–13.

34. Ibid., 13.

35. Paul Lashmar, *Spy Flights of the Cold War* (Gloucestershire: Sutton Limited, 1996), 34–46. For a detailed history of aerial reconnaissance, see Peter Mead, *The Eye in the Air: History of Air Observation and Reconnaissance for the Army 1785–1945* (London: Her Majesty's Stationary Office, 1983).

36. Hays, "Struggling toward Space Doctrine," 66–70; Sheehan, *A Fiery Peace,* 49–151.

37. Robert L. Perry, *Origins of the USAF Space Program,* AFSC Historical Publication Series, 62, 24, 10, George Washington University, National Security Archive [hereafter GWUNSA], Microfilm Collection: The Militarization of Space 1945–1991, Document 00313, 21.

38. Ibid.

39. Air Force ROTC, *Fundamentals of Aerospace Weapons Systems* (Maxwell Air Force Base: Air University, 1961), 274–76; Sheehan, *A Fiery Peace,* 136–42.

40. Walter J. Boyne, *Beyond the Wild Blue: A History of the U.S. Air Force 1947–1997* (New York: St. Martin's Press, 1997), 103–6.

41. Scott D. Sagan, *Moving Targets: Nuclear Strategy and National Security* (Princeton: Princeton University Press, 1989), 15–18.

42. The use of the term "aerospace" began in the US Air Force in 1958. The term was developed to advocate the idea that air and space were a single operational continuum, and that this continuum was the domain of USAF.

43. Perry, *Origins of the USAF Space Program,* 14; of interesting note is that the wording in the original document varies from Perry's quotation. In the document the staff at RAND stated: "In making the decision as to whether or not to undertake construction of such a craft now, it is not inappropriate to view our present situation as similar to that in airplanes prior to the flight of the Wright brothers. We can see no more clearly all the utility and implications of spaceships than the Wright brothers could see fleets of B-29's bombing Japan and air transports circling the globe (Douglas Aircraft Corporation, *Preliminary Design,* 1).

44. Ira C. Eaker to W. Stuart Symington, May 13, 1946, "A Projectile to the Moon," HSTPL, Papers of Stuart Symington, Box 4, File: Correspondence File- 1946–50 Declassified documents, 1.

45. Ibid.

46. The lunar probe discussed never materialized. Roughly a decade later, on September 12, 1959, the Soviet Union successfully launched its lunar probe LUNA 2, achieving the first man-made impact on the surface of the moon.

47. Lee Bowen, *The Threshold of Space: The Air Force in the National Space Program, 1945–1959* (Washington D.C.: USAF Historical Division Liaison Office), 1960, GWUNSA, Microfilm Collection: The Militarization of Space, Document # 00314, 5.

48. Perry, *Origins of the USAF Space Program,* 23.

49. Ibid.

50. RAND Corporation, "Conference on Methods for Studying the Psychological Effects of Unconventional Weapons," Santa Monica, California, January 26–28, 1949, GWUNSA, Microfilm Collection: Militarization of Space, Document 00593, 1. Participants in the conference included Bernard Brodie, Ansley Coale, Irving L. Janis, Harold D. Lasswell, Louis Ridenour, Leo C. Rosten, Frederick F. Stephan, F. R. Collbohm, W. P. Davison, H. Goldhamer, J. M. Goldsen, D. Griggs, C. Hitch, A. Kaplan, P. Kecskemeti, N. Leites, J.E. Lipp, R. Nichols, E. Paxon, E. Plesset, P. Selznick, H. Speiel, J. D. Williams, and L. Young.

51. Ibid.

52. Ibid., 84–85.

53. The process and ideas streaming from this roundtable debate encompassed issues such as the significance of painting the satellite black, the encryption integrity of using televi-

ing of the NSC" [hereafter 310th NSC Meeting], DDEPL, 310th Meeting of the NSC, Box 8, PPDDE 1953–1961, Ann Whitman File, NSC Series, 2; Harford, *Korolev,* 123.

20. 310th NSC Meeting, 7.

21. Ibid., 4.

22. S. Everett Gleason, memorandum, May 10, 1957, "Discussion at the 322nd NSC Meeting" [hereafter 322nd NSC Meeting] DDEPL, 322nd Meeting of NSC, Box 8, PPDDE, Ann Whitman File, NSC Series, 4.

23. S. Everett Gleason, memorandum, October 11, 1957, "Discussion at the 339th Meeting of the NSC" [hereafter 339th NSC Meeting], DDEPL, 339th Meeting of NSC, Box 9, PPDDE, Ann Whitman File, NSC Series, 4; 310th NSC Meeting, 3.

24. United States Information Agency (USIA), *West European Climate of Opinion on the Eve of the Paris Summit Conference,* April 1960, DDEPL, U.S. President's Commission on Information Activity Abroad (PCIAA) # 23, File 5, Box 25, PCIAA Records: 1959–1961, 2; USIA, *U.S. and Soviet Science and Technology in Foreign Public Opinion,* April 21, 1960, DDEL, PCIAA # 23, File 6, Box 22, PCIAA: Records 1959–1961, 4–14.

25. Philip Taubman, *Secret Empire: Eisenhower, the CIA, and the Hidden Story of America's Space Espionage* (New York: Simon and Schuster, 2003), 68–70; Richard A. Hand, Bonnie Houchen, and Lou Larson, eds., *Space Handbook: A War Fighter's Guide to Space, Volume 1* (Maxwell Air Force Base: Air University Press, 1993), 4; Robert Perry, *Origins of the USAF Space Program, 1945–1956* (United States Air Force, Space Systems Division, 1961), 33.

26. Ibid.

27. Ibid., 35.

28. Dwayne A. Day, John M. Logsdon, and Brian Latell, eds. *Eye in the Sky: The Story of the Corona Spy Satellites* (Washington D.C: Smithsonian Institution Press, 1998), 31.

29. Cabell to Schriever, "RAND Satellite Reconnaissance Vehicle," 1.

30. Day, Logsdon, and Latell, *Eye in the Sky,* 31.

31. Taubman, *Secret Empire,* 68; Day, Logsdon, and Latell, *Eye in the Sky,* 31.

32. Taubman, *Secret Empire,* 69.

33. Cabell to Schriever, "RAND Satellite Reconnaissance Vehicle," 1.

34. UASF Space Systems Division, *History of Space Systems Division: July-December 1962 Volume 11* NASAHO, MIDAS, 9; R. Cargill Hall "Missile Defense Alarm: The Genesis of Space-based Infrared Warning" in *Quest* 7 (Spring, 1999), 5.

35. Hall, "Missile Defense Alarm," 5.

36. A. J. Goodpaster, memorandum to, October 8, 1957, "Conference with the President," DDEPL, October 1957 Staff Notes (2), Box 27, PPDDE, Ann Whitman File 1953–1961, Dwight D. Eisenhower Diary Series [hereafter DDEDS], 1; Staff Notes no. 210. October 7, 1957, Toner Notes, Box 7, PPDDE, Ann Whitman File, DDEDS, 2.

37. Press and Radio Conference #123 [hereafter P&RC #123], October 9, 1957, DDEPL, Ann Whitman File, Press Conference Series, Box 6, File: Press Conference 10/09/1957, 7.

38. A. J. Goodpaster, memorandum to, October 8, 1957, 1.

39. 322nd NSC Meeting, 4; P&RC #123, 3.

40. Marion W. Boggs, memorandum, July 30, 1959, "Discussion at the 415th Meeting of the NSC" [hereafter 415th NSC Meeting], DDEPL, 415th Meeting of NSC, Box 11, PPDDE, Ann Whitman File, NSC Series, 7.

41. National Security Council, *Preliminary U.S. Policy on Outer Space* [hereafter NSC-5814/1], August 18, 1958, DDEPL, Outer Space (4), Box 38, WHO NSCSP 1948–1961, Disaster File, 2.

42. Ibid., 2–3.

43. Everett C. Dolman, *Astropolitik: Classical Geopolitics in the Space Age,* (London: Frank Cass, 2002), 8.

44. NSC-5814/1, 7.

45. Ibid., 8.

46. Thomas D. White, "At the Dawn of the Space Age," *Air Power Historian* 15 (January 1958): 15, reprint of address before the National Press Club, Washington D.C., November 29, 1957.

47. Ibid., 17; David N. Spires, *Beyond Horizons: A Half Century of Air Force Space Leadership* (Maxwell Air Force Base: Air University Press, 1998), 54.

48. White, "At the Dawn of the Space Age," 16.

49. Ibid., 18.

50. Alexander de Seversky, "Air Power, Missiles, and National Surival," *Air Power Historian* 5 (January 1, 1958): 22.

51. Spires, *Beyond Horizons,* 54.

52. Homer Boushey, "The USAF and Astronautical Development," *Air Power Historian* 5 (April 1958), 118, 123, from an address given on January 28, 1958 before the Aero Club of Washington D.C.

53. Ibid., 121.

54. Ibid., 123.

55. Boushey, "The USAF and Astronautical Development," 123; Spires, *Beyond Horizons,* 55; United Nations, *The United Nations and Disarmament* (New York: United Nations, 1970), 174; Arms Control and Disarmament Agency, "New Conference Remarks of John Foster Dulles Regarding Outer Space," January 16, 1958, in *Documents on Disarmament 1945–1959* (Washington D.C.: Department of State, 1960), 941–42.

56. Dwayne A. Day, "Invitation to Struggle: The History of Civilian Military Relations in Space," in *Exploring the Unknown: Selected Documents in the History of the U.S. Civilian Space Program, Volume II: External Relations,* John M. Logsdon, ed. (Washington D.C.: NASA History Office, 1996), 233; Lee Bowen, *The Threshold of Space: The Air Force in the National Space Program, 1945–1959* (Washington D.C.: USAF Historical Division Liaison Office, 1960), 13.

57. Day, "Invitation to Struggle," 247.

58. 415th NSC Meeting, 8.

59. Robert Piper, *The Space Systems Division—Background (October 1957–June 1962* (Air Force Systems Command, February 1963), GWUNSA, Militarization of Space Collection, Document 00304, 9.

60. "Briefing of Army Satellite Program," November 10, 1957, GWUNSA, Militarization of Space Collection, Document 00604, 1. The name of the presenter is not listed in the document, and attempts to identify the army's representative have been unsuccessful.

61. 339th NSC Meeting, 7; 322nd NSC Meeting, 4.

62. J. D. Hunley, ed., *The Birth of NASA: The Diary of T. Keith Glennan* (Washington D.C., NASA History Office, 1993), ix.

63. Ibid., *Birth of NASA,* 2.

64. Robert W. Buchheim, et al., *Space Handbook: Astronautics and its Applications* (Santa Monica: RAND Corporation, 1958), foreword.

65. Ibid., 6.

66. This does not assume that the military space programs had been consciously designed to provide specific civilian spinoffs. Rather, it seems that the missions (weather data collection, communication, and navigation) first identified with military satellites could easily be adjusted to fit a future nonmilitary market.

67. White House Office, memorandum for J. R. Killian, February 21, 1958, "Preliminary Observations on the Organization for the Exploitation of Outer Space," GWUNSA, Militarization of Space Collection, Document 500, 1.

68. Ibid.

69. Ibid.

70. "Johnson Linked to the Space Arms Ban," *Star Telegram,* November 11, 1958, Lyndon Baines Johnson Presidential Library [hereafter LBJPL], Outer Space Speeches, United States Senate, 1949–1961, Box 359, 1.

71. Ibid.

72. "Space Conquest May Bring World Peace, Johnson Says," *Dallas Morning News,* November 12, 1958, LBJL, Outer Space Speeches, United States Senate, 1949–1961, Box 359, section 1 page 8.

73. Ibid.

74. Jack Manno, *Arming the Heavens: The Hidden Military Agenda for Space 1945–1995* (New York: Dodd, Mead, 1984), 5. Manno's work is the most glaring example of this position. This historiographic school emerged after March 1983, when President Ronald Reagan made his "Strategic Defense Initiative speech." The emergence of this school evolved out of a fear that Reagan's proposal would lead to a proliferation of space-based weapons, and a renewed struggle for control of the high ground. Hence, these scholars strove to prove that the United States had a history of seeking to arm space.

75. 415th NSC Meeting, 8.

76. "The Baruch Plan: Statement by the United States Representative to the United Nations Atomic Energy Commission, June 14, 1946," in *Documents on Disarmament 1945–1959, Volume I, 1945–1956* (Washington D.C.: Arms Control and Disarmament Agency (ACDA), Department of State, 1960), 9.

77. "News Conference Remarks by Secretary of State Dulles Regarding Outer Space, January 16, 1958," in *Documents on Disarmament 1945–1959, Volume II, 1957–1959* (Washington D.C.: ACDA, Department of State, 1960), 942.

78. "Western Power Paper Submitted to the Disarmament Subcommittee: Proposals for Partial Measures on Disarmament, August 29, 1957," in *Documents on Disarmament 1945–1959, Volume II, 1957–1959* (Washington D.C.: ACDA, Department of State, 1960), 871–74.

79. A discussion of the Antarctic Treaty can be found in *Arms Control and Disarmament Agreements: Texts and Histories of the Negotiations* (Washington D.C.: ACDA, 1996).

80. Bowen, *The Threshold of Space,* 37.

81. Ibid. Each of these five systems had a variety of subsystems attached to the larger program. System 609 included BRATS and the Aerial Survey and Target locating systems. System 447 was the X-15 and a planned follow-on vehicle called "advanced hypersonic research aircraft." System 464 provided development such as "boost glide tactical, boost glide intercept, satellite intercept, global reconnaissance, and global bomber" for the Dyna-Soar program. WS-117L involved research on "advanced reconnaissance satellite, recoverable photo capsule, manned strategic station, 24 hour reconnaissance, and strategic communication station." Finally, 499 included "manned variable trajectory and test vehicle, nuclear rocket test, ion propulsion system, lunar transport, and manned lunar base."

82. Ibid., 27.

83. Ibid. The inflatable sphere assigned to the US Army later become the passive communications satellite known as Project Echo. The Echo satellite acted as a reflector to bounce radio signals to distant points on the globe.

84. "Midas—America's First IR Ballistic Missile Early Warning Satellite Program—

Declassified," news release, Office of Public Affairs: Space and Missile Center, United States Air Force, March 10, 1999, NASA History Office [hereafter NASAHO], MIDAS, 1.

85. *The Report to the President by the Technological Capabilities Panel of the Science Advisory Committee* [hereafter Killian Report], February 14, 1955, DDEPL, Technological Capabilities Panel (1), Box 16, WHO, Office of Staff Secretary 1952–1961, Subject Series, Alphabetical Subseries, 114–23.

86. "Midas—America's First IR Ballistic Missile Early Warning Satellite Program," 1.

87. Hall, "Missile Defense Alarm," 7.

88. Ibid.

89. Thomas A. Power, memorandum to Thomas D. White, May 14, 1959, NASAHO, SAMOS Satellites-SAC, 1.

90. Cabell to Schriever, "RAND Satellite Vehicle," 1; Merton E. Davies and William R. Harris, *RAND's Role in the Evolution of Balloon and Satellite Observation Systems and Related U.S. Space Technology* (Santa Monica: RAND Corporation, 1988), 3–63

91. Greer, "Corona," 5.

92. Ibid.; Day, Logsdon, and Latell, *Eye in the Sky,* 107–18; and Taubman, *Secret Empire,* 245.

93. Michael Bechloss, *May Day: Eisenhower, Khrushchev, and the U-2 Affair* (New York: Harper Row, 1986), 363–64; Taubman, *Secret Empire,* 315.

94. Bechloss, *May Day,* 23–27, 355–63, 376–77; Taubman, *Secret Empire,* 305–9.

95. "Army Contributes to Development of TIROS Satellite Presented to Smithsonian," news release, April 1, 1965, U.S. Army Material Command, NASAHO, Project TIROS, 1.

96. NASA, *TIROS: Television Infra-Red Observation Satellite* (Washington D.C.: NASA, 1961), NASAHO, TIROS, 1.

97. NASA, "Project TIROS: Information and Guideline Meeting," October 21, 1959, NASAHO, TIROS, 1.

98. The US government overtly acknowledged the Corona program in February 1995, when the CIA declassified the photographs and history of the program. See Kevin Ruffner, ed., *Corona: America's First Satellite Program* (Washington D.C.: Center for the Study of Intelligence, 1995), xi.

99. Robert Qualkinbush, "TRANSIT: The U.S. Navy Pioneers Satellite Navigation," *Journal of British Interplanetary Society* 50 (1997): 404.

100. Applied Physics Laboratory, "TRANSIT: 1958 to 1996," undated, NASAHO, TRANSIT, 1.

101. APL, Public Information Office, "Navigation Satellite Travels 5.8 Billion Miles/ Celebrates 20 years in Orbit," May 4, 1987, NASAHO, TRANSIT, 2–3.

102. While the history of the development of civilian communication satellites is beyond the scope of this project, a solid treatment of the topic can be found in David Whalen, *The Origins of Satellite Communication, 1945–1965* (Washington D.C.: Smithsonian Institution Press, 2002).

103. Joe Stein, "Proposed ARPA Release," August 4, 1958, NASA HQ, ECHO, 1; NASA, "ECHO I," Undated, NASAHO, ECHO, 2–3.

104. "Military Space Projects, March–April–May 1960," August 16, 1960, Washington D.C.: Department of Defense, Office of the Director of Defense Research and Engineering, GWUNSA, Militarization of Space Collection, Document 00321, 55.

105. Whalen, *The Origins of Satellite Communications,* 2–3.

106. Trevor Gardner, "How We Fell Behind in Guided Missiles," *Air Power Historian* 5 (January 1958), 4; White, "At the Dawn of the Space Age," 16; Boushey, "The USAF and Astronautical Development," 118.

107. Paul B. Stares, *The Militarization of Space, U.S. Policy 1945–1984* (Ithaca: Cornell University Press, 1985), 52. Note that this quotation comes from an interview with a DoD official who wished to remain unnamed. The difference between an inspector and an interceptor satellite is that the inspector system possessed a capability to discriminate other satellites before a decision was made on intercepting. The interceptor satellite originally had no inspection capability.

108. George B. Kistiakowsky, *A Scientist at the White House* (Cambridge: Harvard University Press, 1976), 246; Stares, *Militarization of Space,* 53.

109. Stares, *Militarization of Space,* 53.

110. As the objective of this book is to detail America's quest to militarize space, while remaining committed to the principle of preserving space as a weapons-free environment, the various ballistic missile defense systems envisioned or studied between 1952 and 1960 will not be thoroughly discussed. Only those systems directly involving the use of space-based interceptors or warning networks will be highlighted. For a history of American BMD programs, see Donald Baucom, *The Origins of SDI, 1944–1983* (Lawrence: University of Kansas Press, 1992).

111. Ibid., 15.

112. Lisa M. Borowski, "Military Priorities in the Cold War: Operation ARGUS, 1957–1959" (master's thesis, Kansas State University, 2000), 1.

113. Baucom, *Origins of SDI,* 16.

Chapter 3

1. "Eisenhower Asks to End Space Race," *Space Daily,* August 9, 1962, NASAHO Eisenhower, 169; "IKE Says Space Spending Too High," *Space Daily,* April 7, 1964, NASAHO, Eisenhower, 37. According to a June 13, 1963, *Chicago Tribune* article entitled "Moon Race Nuts," former president Eisenhower was quoted as saying "anyone who would spend $40 billion dollars in a race to the moon just for national prestige is 'nuts.'" "Moon Race Nuts," *Chicago Tribune,* June 13, 1963, NASAHO, White House, Eisenhower, Post-White House.

2. Barry Goldwater, "America in Space," *Air Power Historian* 10 (April 1963): 43; Marvin W. McFarland, "Men of Space," *Air Power Historian* 10 (October 1963): 127; Curtis E. LeMay, "Our National Position in the World from a Military Point of View," *Air Power Historian* 9 (July 1962): 138.

3. Transcript of President Kennedy's news conference, September 13, 1962, NASAHO, Kennedy, 15a. Transcript of Kennedy Speech on November 18, 1961, before the Democratic Party of California, NASAHO, Kennedy, 15a.

4. United States Information Agency, "West European Opinion Trends on U.S. and Soviet Strength," February 1963, John F. Kennedy Presidential Library [hereafter JFKPL], Papers of President Kennedy [hereafter PPJFK], National Security File, Department and Agency, Box 290, File: USIA General 1/61–6/61, 1–8; Central Intelligence Agency, "A Brief Look at the Soviet Space Program," October 1, 1963, JFKPL, PPJFK, National Security File, Subjects, Box 308, File: Space Activities General 10/63–11/63, 1.

5. Theodore Brassord, Jr. letter to Senator John F. Kennedy, January 8, 1960, JFKPL, PPJFK, Pre-Presidential, Box 747, File: Space, 2.

6. William Everdall letter to Senator John F. Kennedy, February 3, 1960, John F. Kennedy Presidential Library, PPJFK, Pre-Presidential, Box 747, File: Space, 1.

7. For an assessment of America's prestige at the end of Eisenhower's tenure, see USIA,

"Western European Climate of Opinion on the Eve of the Paris Summit," April 1960, 2.; and USIA, "U.S. and Soviet Science and Technology in Foreign Public Opinion," April 21, 1960, 4–14.

8. John F. Kennedy letter to William Everdall, February 19, 1960, JFKPL, PPJFK, Pre-Presidential, Box 747, File: Space, 1–2.

9. In April 1964, Eisenhower was quoted in the *Saturday Evening Post* as calling Kennedy's space program a "stunt," which Eisenhower believed to be funded "all out of proportion." See "Ike Says Space Spending Too High," *Space Daily*, April 7, 1964, NASAHO, Eisenhower, Post-White House, 37.

10. Donald H. Menzel, memorandum to Senator John F. Kennedy, "Space and Science Programs of the United States," September 7, 1960, JFKPL, PPJFK, Pre-Presidential, Box 747, File: Space, 1.

11. Selected Statements of President Kennedy on Defense Topics, December 1957–August 1, 1962, "Ground Support Equipment," April–May 1960, NASAHO, White House—President John F. Kennedy, File: Kennedy Statements on Defense, 109. Kennedy stated that the Soviet Union's space program created "apprehension on the part of the non-communist nations that reliance on the United States to preserve the balance of world power [is] unwarranted. The Soviet space surge is of enormous value in the Russian effort in creating a world image of an inevitable and therefore unbeatable world system."

12. Ibid.

13. Ibid., 15a.

14. Excerpts from the Special Message to Congress by the President of the United States, May 25, 1961, NASAHO, White House—President John F. Kennedy, File: Kennedy Speeches, 1–2.

15. John F. Kennedy, transcript of speech on September 12, 1962, NASAHO, White House—Presidents, Kennedy, File: Kennedy Speeches, 2–3.

16. Ibid.

17. National Reconnaissance Office, Information Management Services, NRO Staff Records, NSAM 156 Committee, Undated, 1.

18. Lyman L. Lemnitzer, draft of address given at the Key Issues Forum, Academy of Music, Philadelphia, October 18, 1961, JFKPL, PPJFK, National Security File, Box 276, File: DoD (B) Joint Chief of Staff, General 1961, 1.

19. Edward W. Murrow, memorandum to Mr. Gilpatrick, March 8, 1961, "U.S. Public Position on Our Defense Capability," JFKPL, PPJFK, National Security File, Box 290, File: USIA, General, 1/61–6/61, 1. Murrow stated "it is basically important to us that people overseas—enemies, allies, and neutrals—think the United States is strong. We should, therefore, do as much as we can to reverse the recent trend of overseas opinion saying the U.S. is weak and to build an image of superior and growing strength."

20. Carl Berger, *The Air Force in Space. Fiscal Year 1961* (USAF Historical Division Liaison Office, April 1966), GWUNSA, Militarization of Space Microfilm Collection, Documents 00323, 1; Carl Berger, *The Air Force in Space, Fiscal Year 1962* (USAF Historical Division Liaison Office, June 1966), GWUNSA, Militarization of Space Microfilm Collection, Document 00327, 5.

21. Berger, *Air Force in Space, Fiscal Year 1962,* 1.

22. Berger, *Air Force in Space, Fiscal Year 1961,* 29.

23. Ibid., 30.

24. Ibid., 32.

25. Dwayne Day, "Invitation to Struggle: The History of Civilian-Military Relations in Space," in *Exploring the Unknown: Selected Documents in the History of the U.S. Civilian Space*

Program, Volume II: External Relationships, John M. Logsdon, Dwayne Day, and Roger D. Launius, eds., (Washington D.C.; NASA History Office, 1996), 233–55.

26. US Air Force, *The Space Plan,* September 1961, JFKPL, PPJFK, National Security File, Box 307, File: Space Activities, General, 1/61–3/61, v.

27. Ibid., 1.

28. Ibid., 5.

29. Ibid., 43; Berger, *Air Force in Space, Fiscal Year 1962,* 19.

30. David N. Spires, *Beyond Horizons: A Half Century of Air Force Space Leadership* (Maxwell Air Force Base: Air University Press, 1998), 104.

31. Carl Berger, *The Air Force in Space, Fiscal Year 1962* (USAF Historical Division Liaison Office, June 1966), GWUNSA, The Militarization of Space Microfilm Collection, Document 00327, 21; Spires, *The Militarization of Space,* 105. For an insightful treatment of the close relationship between the DoD and NASA, see Mark Erickson, *Into the Unknown Together: The DoD, NASA, and Early Spaceflight* (Maxwell Air Force Base, Alabama: Air University Press, 2005).

32. Curtis Peebles, "The Origins of the U.S. Space Shuttle," *Spaceflight* 21 (November 19790): 435; R. F. Houchin, "Doctrine and Dyna-Soar: Origins of USAF Manned Military Spaceflight," 48th International Astronautical Congress, Turi , Italy, October 6–10, 1997, NASAHO, File: DYNA-SOAR, 1; Robert Godwin, ed. *Dyna-Soar: Hypersonic Strategic Weapon* (Ontario: Apogee, 2003), 8–9.

33. Peebles, "Origins of U.S. Space Shuttle," 435; Houchin, "Doctrine and Dyna-Soar," 1.

34. Peebles, "Origins of U.S. Space Shuttle," 436; Houchin, "Doctrine and Dyna-Soar," 3.

35. Peebles, "Origins of U.S. Space Shuttle," 436.

36. Robert S. McNamara, transcript of Department of Defense news briefing on December 10, 1963, in the Pentagon, NASAHO, File: Dyna-Soar, 1.

37. Boeing Company, "This is DYNA-SOAR," undated, NASAHO, File: Dyna-Soar, 6–11; Peebles, "Origins of U.S. Space Shuttle," 436.

38. James Baar, "Push to Speed Up Dyna-Soar, but Administration Is Still Unconvinced of Requirements," *Missiles and Rockets,* August 14, 1961, 14.

39. Melvin R. Laird, copy of transcript of remarks for the *Congressional Record,* April 17, 1962, NASAHO, File: Dyna-Soar, 30.

40. "DYNA-SOAR Designated X-20 by the Air Force," news release, June 26, 1962, Department of Defense, Office of Public Affairs, NASAHO, Dyna-Soar, 1; Godwin, ed. *Dyna-Soar,* 248.

41. Joseph V. Charyk, transcript of remarks made before the Air Force Association in Las Vegas, Nevada, September 20, 1962, NASAHO, File: Dyna-Soar, 1.

42. Department of Defense, "Air Force to Develop Manned Orbital Laboratory," press release, December 10, 1963, NASAHO, File: Dyna-Soar, 1.

43. Ibid.

44. "Decision on Manned Orbiting Laboratory and Related Matters," *NASA News,* January 25, 1965, NASAHO, File: MOL, 1–6; "NASA MOL In-House Position Paper," *Space Daily,* January 6, 1964, NASAHO, File: MOL, 20.

45. Jerome Wiesner, memorandum to President Kennedy, February 20, 1961, untitled, JFKPL, PPJFK, National Security Files, Subjects, Box 307, File: Space Activities General 1/61–3/61, 1–2.

46. Ibid.

47. "Recommendations for Our National Space Program: Changes, Policies, and Goals," undated, JFKPL, PPJFK, National Security Files, Subjects, Box 30, File: Space Activities 4/61–6/61, 19.

48. Ibid.

49. John F. Kennedy, transcript of speech made at American University, June 10, 1963, NASAHO, White House–Presidents, John F. Kennedy, Defense Statement, File: Kennedy Comments on Defense August1962–July 1963, 225.

50. John F. Kennedy, transcript of speech before the United Nations, September 25, 1961, *Selected Statements of President Kennedy on Defense Topics, December 1957–August 1, 1962,* NASAHO, Presidents, John F. Kennedy, File: Kennedy–Statements on Defense, 1957–August 1962, 213.

51. John F. Kennedy, letter to Congressman Albert Thomas, September 23, 1963, JFKPL, PPJFK, National Security File, Subjects, Box 308, File: Space Activities General 7/63–9/63, 1; James Webb, "NASA New Release, Transcript of Speech at the Annual Meeting of the Texas Mid-Continent Oil and Gas Association," Houston, Texas, September 25, 1963, JFKPL, PPJFK, National Security File, Subjects, Box 308, File Space Activities General 7/63–9/63, 1. Webb noted in his speech, "In itself, cooperation in space activities between the great antagonists of the Cold War is a thrilling prospect."

52. S. Everett Gleason, memorandum, "Discussion at the 339th Meeting of the National Security Council," October 11, 1957, DDEPL, DDE Papers as President, NSC Series, Box 9, File: 339th Meeting of the NSC, 4.

53. Charles S. Sampson and Glenn W. LaFantasie, eds., *Foreign Relations of the United States, 1961–1963, Volume VI, Kennedy-Khrushchev Exchanges* (Washington D.C.: Government Printing Office, 1996), ix–xii.

54. Nikita Khrushchev, letter to Kennedy, reprinted in *New York Times,* March 22, 1962, JFKPL, PPJFK, National Security File, Subjects, Box 308, File: Space Activities US-USSR Cooperation 1961–1963, 1–2.

55. Roswell Gilpatric, transcript of remarks before the San Francisco Chamber of Commerce, May 16, 1962, JKFPL, PPJFK, National Security File, Subjects, Box 307, File: Space Activities General 4/62–5/62, 5.

56. Roswell Gilpatric, transcript of remarks at AFSC Management Conference, Monterey, California, May 2, 1963, in news release, Department of Defense, Office of Public Affairs, JFKPL, PPJFK, National Security File, Carl Kaysen Series, Box 377, File: Space Policy General 1961–1963, 8.

57. Gilpatric, remarks before San Francisco Chamber of Commerce, 5.

58. Outgoing telegram: Department of State to U.S. Mission Geneva, June 22, 1962, JFKPL, PPJFK, National Security File, Subjects, Box 308, File: Space Activities Cables 1/61–7/62, 3.

59. "The President's Policy Statement on Outer Space," draft, 6/14/1962, JFKPL, PPJFK, National Security File, Subjects, Box 307, File: Space Activities General 6/62–7/62, 1.

60. Everett C. Dolman, *Astropolitik: Classical Geopolitics in the Space Age,* (London: Frank Cass, 2002), 123.

61. John F. Kennedy, transcript of news conference, June 14, 1962, NASAHO, President John F. Kennedy, File Kennedy–Statements on Defense, 1957–August of 1962, 207.

62. Jerome Wiesner, memorandum to the president, August 9, 1962, JFKPL, PPJFK, National Security File, Department and Agencies, Box 284, File: Office of Science and Technology, General 1961–1962, 1.

63. Ibid.

64. Ibid.

65. Ibid., 6.

66. William C. Foster, memorandum for the Committee of Principals, "U.S. Approach to a Separate Arms Control Measure for Outer Space," September 15, 1962, JFKPL, Na-

tional Security Files, Carl Kaysen Series, Box 377, File: Space Policy, Subjects, Satellite 9/62–12/62, 1.

67. Ibid.

68. "Recommendations Respecting U.S. Approach to a Separate Arms Control Measure for Outer Space," attached to ibid.

69. Ibid.

70. U. Alexis Johnson, memorandum for Mr. Nitze, Dr. Charyk, Dr. Wiesner, Mr. Kaysen, Dr. Scoville, Mr. Fisher, Dr. Seamans, and Mr. Murrow, "Position and Contingency Papers Pursuant to NSAM 183," September 21, 1962, JFKPL, PPJFK, National Security File, Carl Kaysen Series, box 377, File: Space Policy—Subjects, Satellites 1963, 1.

71. A year later, on September 21, 1963, Soviet Foreign Minister Gromyko said before the UN General Assembly that the Soviet Union "deemed it necessary to take steps to prevent the spread of an arms race in outer space." Specifically Gromyko mentioned the need to ban "nuclear and other weapons of mass destruction." United Nations, *The United Nations and Disarmament* (New York: United Nations, 1970), 176.

72. "U.S. Outer Space Policy in the UN General Assembly," Attachment 1, "Position and Contingency Papers Pursuant to NSAM 183," September 21, 1962, 1.

73. Ibid.

74. Ibid., 3.

75. Ibid., 4.

76. "Definition of Peaceful Uses of Outer Space," Attachment 2, "Position and Contingency Papers Pursuant to NSAM 183," 1.

77. J. M. Goldsen, *Military Space Activities and the Cold War* (Santa Monica: RAND Corporation, December 1962), JFKPL, PPJFK, National Security File, Subjects, Box 307, File: Space Activities, General 1/63–5/63, v.

78. Ibid., 5; V. D. Sokolovsky, *Soviet Military Strategy,* 3rd ed., Harriet Fast Scott, ed. (New York: Crane, Russak, 1968), 84–89.

79. In 1946, RAND engineers quickly dismissed using satellites as bombing platforms based upon the lack of an adequate booster to lift large and heavy bombardment satellites into orbit. See RAND Corporation, *The Preliminary Design of an Experimental World-Circling Spaceship* (Santa Monica: RAND Corporation, 1946), 1–17.

80. Pavel Podvig, ed., *Russian Strategic Nuclear Forces* (Cambridge: MIT Press, 2001), 196–200; Cecil Brownlow, "Soviets Prepare Space Weapons for 1968," *Aviation Week and Space Technology* 108 (November 13, 1967), 30–31.

81. U. Alexis Johnson, Attachment 1, "Guidelines for Public Statements on Unilateral U.S. Capabilities to Monitor a Ban on Weapons in Space," memorandum for Mr. Nitze, Dr. Charyk, Dr. Seamans, Dr. Wiesner, Mr. Kaysen, Dr. Scoville, and Mr. Fisher, "Further Action. Under NSAM 192," February 19, 1963, JFKPL, PPJFK, National Security File, Carl Kaysen Series, File: Space Policy, Subjects, Satellites 9/62–12/62, 1.

82. Ibid., 2.

83. Ibid., attachment: "Contingency Plan for US Reaction to Soviet Placing of a Nuclear Weapon in Space," 3–4; Dean Rusk, memorandum to the president, May 8, 1963, "U.S. Reaction to Soviet Placing of a Nuclear Weapon in Space," JFKPL, PPJFK, National Security File, Carl Kaysen Series, Box 377, File: Space Policy–Subjects, Satellites 1963, 1–3.

84. Department of State Airgram from Embassy Prague to Department of State, "Objection to Soviet Space Exhibit," May 24, 1963, JFKPL, PPJFK, National Security Files, Subjects, Box 308, File: Space Activities Cables 8/62–11/68, 1.

85. "A U.S.-Soviet Arrangement concerning the Placing in Orbit of Weapons of Mass Destruction," attachment to memorandum to Committee of Deputies from Adrian S. Fisher,

"Proposed U.S.-Soviet Arrangement Concerning the Placing in Orbit of Weapons of Mass Destruction," October 1, 1963, JFKPL, PPJFK, National Security File, Subjects, Box 308, File: Space Activities Bombs in Orbit 9/63–11/63, 3.

86. Ibid., 14.

87. Clayton K. S. Chun, *Shooting Down a "Star": Program 437, the US Nuclear ASAT System and the Present-Day Copycat Killers* (Maxwell Air Force Base: Air University Press, 2000), 2.

88. Jack Manno, *Arming the Heavens: The Hidden Military Agenda for Space 1945–1995* (New York: Dodd, Mead, 1984), 84; Chun, *Shooting Down a Star,* 437, 4. The satellites were American and British and the three were damaged; the status of an additional four satellites was not commented on.

89. Lisa M. Borowski, "Military Priorities in the Cold War: Operation ARGUS, 1957–1959" (master's thesis, Kansas State University, 2000), 112; Chun, *Shooting Down a Star,* 4.

90. McGeorge Bundy, National Security Action Memorandum No. 258 to Secretary of Defense, Secretary of Commerce, Administrator of NASA, Director Bureau of the Budget, and Director Office of Emergency Planning, August 6, 1963, "Assignment of Highest National Priority to Program 437," JFKPL, PPJFK, National Security File, Subjects, Box 308, File: Space Activities General 7/63–9/63, 1.

91. Paul B. Stares, *The Militarization of Space, U.S. Policy 1945–1984* (Ithaca: Cornell University Press, 1985), 72; Chun, *Shooting Down a Star,* 8. A complete discussion of the history of Nike-Zeus and its follow-on, Nike-X, can be found in James S. Walker, Francis Martin, and Sharon S. Watkins, *Strategic Defense: Four Decades of Progress* (U.S. Army Space and Strategic Defense Command: Historical Office, 1995). Interestingly, Walker, Martin, and Watkins make no mention of the army's decision to add an ASAT role to the Nike-Zeus system.

92. Quoted in Stares, *Militarization of Space,* 73. The original source: US Congress, House Committee on Science and Astronautics, *Military Astronautics,* 87th Congress, 1st Session (1961): 11.

93. Stares, *Militarization of Space,* 117; Chun, *Shooting Down a Star,* 8–9.

94. Chun, *Shooting Down a Star,* 9.

95. While the ASAT programs supported under Kennedy encompassed the possibility of US missiles intercepting USSR satellites in space, the administration held that the interception medium did not violate their pledge against weapons in space. The basing mode, rather than the interception point, became the primary determinant of their policy. Therefore, Kennedy and his administration held that ASATs were compatible with their policies because the ASAT interceptors were ground-based and literally had the power to keep weapons out of space.

Chapter 4

1. Robert Dallek, "Johnson, Project Apollo, and the Politics of Space Program Planning," in *Spaceflight and the Myth of Presidential Leadership,* Roger D. Launius and Howard E. McCurdy, eds. (Urbana: University of Illinois Press, 1997), 68–88.

2. George Reedy, memorandum to Senator Johnson, January 24, 1952, LBJPL, Senate Papers 1949–1961, Committee on Armed Services, Box 346, File: Preparedness Subcommittee Correspondence and Memorandum 82nd Congress, 1951–1952, 1.

3. "Outline of Testimony Given before Senator Johnson's Subcommittee on Soviet

Guided Missiles and Their Related Capabilities, by DCI Allen W. Dulles and Dr. Herbert Scoville," undated, LBJPL, Senate Papers, Committee on Armed Services, Box 355, File: Preparedness Subcommittee on Satellites and Missile Program Hearings, 2.

4. Howard E. Newell, "U.S., Russian Space Efforts Compared," *Aviation Week and Space Technology,* December 21, 1959, 36; United States Information Agency (USIA), "West European Opinion Trends on U.S. and Soviet Strength," February 1963, JFKPL, PPJFK, National Security Files, Department and Agencies Box 290, File: USIA General 2/63-3-/63, 1–8; USIA, "West European Climate of Opinion on the Eve of the Paris Summit Conference," April 1960, 2; USIA, "U.S. and Soviet Science and Technology in Foreign Public Opinion," April 1960, 4–14.

5. Johnson's remarks, undated, untitled, LBJPL, Senate Papers 1949–1961, Committee on Armed Services, Box 355, File: Preparedness Subcommittee on Satellite and Missile Programs Hearings, 4.

6. Solis Horwitz, memorandum to Senator Johnson, October 11, 1957, "Missile Inquiry," LBJPL, Senate Papers 1949–1961, Committee on Armed Services, Box 355, File: Preparedness Subcommittee on Satellite and Missile Program Hearings, 1.

7. Ibid., 2.

8. Ibid.

9. Walter A. McDougall, *The Heavens and the Earth: A Political History of the Space Age* (Baltimore: Johns Hopkins University Press, 1985), 195–209.

10. Mrs. W. A. Balendock, letter to Senator Johnson, October 15, 1957, LBJPL, Senate Papers 1949–1961, Committee on Armed Services, Box 355, File: Satellite (2 of 2), 1.

11. Paul Dickson, *Sputnik: The Shock of the Century* (New York: Walker and Company, 2001), 4.

12. Ibid., 62.

13. Robert A. Divine, *The Sputnik Challenge: Eisenhower's Response to the Soviet Satellite* (New York: Oxford University Press, 1993), 62.

14. Ibid.

15. Press release from the office of Senator Style Bridges, November 7, 1957, LBJPL, United States Senate 1949–1961, Committee on Armed Services, Box 356, File: Other People's Statements on Satellite and Missile Program, 2.

16. Divine, *Sputnik Challenge,* 64.

17. McDougall, *Heavens and Earth,* 148.

18. "Report of Special Senate Committee on Space and Astronautics on S. 3609," undated, LBJPL, United States Senate 1949–1961, Committee on Armed Services, Special Committee on Space and Astronautics, Box 357, File: Special Committee on Space and Astronautics [S.3609], 2–3.

19. "Johnson Linked to Space Arms Ban," *Star Telegram,* November 11, 1958, LBJPL, United States Senate 1949–1961, Committee on Aeronautics and Space Sciences Box 359, File: Outer Space Speeches, 1.

20. McDougall, *Heavens and Earth,* 398.

21. Lyndon Baines Johnson, "The Space Age and the Engineer," *American Engineer* 28 (August 1958): 14.

22. Marion Boggs, memorandum, July 30, 1959, "Discussion at the 415th Meeting of the NSC," DDEPL, Papers of the President 1953–1961, Ann Whitman File, NSC Series, Box 11, File: 415th Meeting of the NSC, 3.

23. "Space Conquest May Bring World Peace, Johnson Says," *Dallas Morning News,* November 12, 1958, Section 1, 8.

24. "Sharing of Space Secrets Proposals," *San Antonio News,* November 12, 1958, 2-B.

25. USIA, "West European Climate," 2; USIA, "U.S. and Soviet Science and Technology in Foreign Public Opinion," 4–14.

26. The layout and information of the National Space Program publication closely parallels the Space Handbook presented by the RAND Corporation in 1958. See DoD and NASA, *National Space Program,* June 3, 1961, LBJPL, Papers of LBJ [hereafter PLBJ], Vice President Security File, Box 15, File: National Space Program.

27. Ibid., 1.

28. Ibid., 20–29.

29. Ibid., 24.

30. Ibid., 40.

31. Kevin C. Ruffer, *Corona: America's First Satellite Program* (Washington DC: Center for the Study of Intelligence, 1995), 1–2; Dwayne A. Day, John M. Logsdon, and Brian Latell, eds. *Eye in the Sky: The Story of the Corona Spy Satellite* (Washington D.C.: Smithsonian University Press, 1998), 51.

32. *National Space Program* (see note 26), 48.

33. Ibid., 52–53 and 56–57.

34. Johns Hopkins University Applied Physics Laboratory, *Transit 1958–1996,* undated, NASAHQ, File: Transit, 1; Robert Qualkinbush, "Transit: The U.S. Navy Pioneers Satellite Navigation," *Journal of the British Interplanetary Society* 50 (1997): 403–7; Geoff Richards, "Transit—The First Navigational Satellite System," *Space Flight* 21 (February 1979): 50–55.

35. *National Space Program* 1961, 56.

36. Robert S. McNamara, memorandum for the vice president, April 21, 1961, "Brief Analysis of Department of Defense Space Program Efforts," NASAHO, White House, President—Lyndon B. Johnson, NASA/White House Correspondence, File: Johnson Correspondence 1960–1962, 2.

37. Ibid.

38. George J. Feldman and Charles S. Sheldon III, "Interim Report on Space Policy," April 24, 1961, LBJPL, PLBJ, Vice Presidential Security File, Box 17, File: Space Program—Secret, 1.

39. Ibid.

40. Lyndon Baines Johnson, memorandum for the president, April 28, 1961, "Evaluation of Space Program," NASAHO, White House, President—LBJ, NASA/White House Correspondence, File: Johnson Correspondence 1960–1962, 1.

41. Ibid., 1–6.

42. "Introductory Statement by the Vice President," August 18, 1961, LBJPL, PLBJ, Vice Presidential Security File, Box 15, File: National Space Program, 2. According to the document, there was a fourth presenter whose name has been sanitized from the record.

43. "Questions," August 18, 1961, LBJPL, PLBJ, Vice Presidential Security File, Box 15, File: National Space Program, 1.

44. Colonel Burris, memorandum to the vice president, April 16, 1962, "Soviet Interest in Space Warfare," NASAHO, White House, Presidents—LBJ, NASA/White House Correspondence, File: Johnson Correspondence, 1960–1962, 1.

45. Robert S. McNamara, "Memorandum for the President," October 28, 1967, LBJPL, National Security File, Country File, Box 231, File: USSR—Fractional Orbital Bombardment System, 1.

46. Barry M. Goldwater, speech before the National Rocket Club, Washington D.C., July 17, 1962, reprinted as "America in Space," *Airpower Historian* 10 (April 1963): 43–44.

47. Ibid., 44.

48. Ibid.

49. Ivan Getting, address to the National Rocket Club, Washington D.C., September 18, 1962, reprinted as "The Space Dichotomy," *Airpower Historian* 9 (October 1962): 226.

50. Alexander de Seversky, "Command of Space," *Airpower Historian* 9 (October 1962): 253.

51. Ibid.

52. John F. Kennedy, memorandum to Robert S. McNamara and James Webb, July 22, 1963, LBJPL, National Security File, Files of Charles E. Johnson, Box 15, File: Military Uses of Space, 1.

53. Charles E. Johnson, memorandum to McGeorge Bundy, July 30, 1963, untitled, LBJPL, National Security File, Files of Charles E. Johnson, Box 15, File: Military Uses of Space, 1.

54. Roswell L. Gilpatric, memorandum to President Kennedy, July 31, 1963, "Reader's Digest Article on Space," LBJPL, National Security File, Files of Charles E. Johnson, Box 15, File: Military Uses of Space, 1.

55. Ibid., 2.

56. George Reedy, memorandum to LBJ, August 1, 1963, LBJPL, PLBJ, Vice Presidential Security File, Box 17, File: Space Program—Secret, 1.

57. The members of the Critical Issues Council were Milton Eisenhower, Elliott V. Bell, Adm. Arleigh A. Burke, Arthur F. Burns, Albert L. Cole, James H. Douglas, Marion B. Folsom, Thomas S. Gates, T. Keith Glennan, Oveta Culp Hobby, Walter H. Judd, Mary P. Lord, Clare Booth Luce, Deane W. Malott, James P. Mitchell, Gen. Lauris Norstad, Don Paarlberg, C. Wrede Petersmeyer, Samuel R. Pierce Jr., Charles S. Rhyne, Raymond J. Saulnier, Lewis L. Strauss, Walter N. Thayer, and Henry C. Wallich; "News from Republican Citizens," May 28, 1964, LBJPL, White House Central Files, Outerspace, File: OS Outer Space 6/20/64–9/21/64, 3–4.

58. Critical Issues Council [hereafter CIC], *America's Space Program: An Appraisal,* May 28, 1964, LBJPL, White House Central Files, Outerspace, File: OS Outer Space 6/20/64–9/21/64, 1.

59. "News from Republican Citizens," 1.

60. CIC, *America's Space Program,* 3.

61. "News from Republican Citizens," 1–2.

62. James Webb, memorandum for Walter Jenkins, June 15, 1964, LBJPL, Whitehouse Central Files, Outer Space, File: OS Outer Space 6/20/64—Webb and Dryden Memorandum from WHCF, OS.

63. Lyndon Baines Johnson, letter to Milton Eisenhower, June 29, 1964, LBJPL, White House Central Files, Outer Space, File: OS Outer Space 6/20/64–9/21/64, 1.

64. Spurgeon Keeny, memorandum to McGeorge Bundy, September 16, 1965, "Public Statement on MOL and Military Use of Space," LBJPL, National Security File, Intelligence File, Box 10, File: Manned Orbiting Laboratory, 1; Lyndon Baines Johnson, Transcript of Press Conference no. 49, August 25, 1965, LBJPL, National Security File, Intelligence File, Box 10, File: Manned Orbiting Laboratory, 1.

65. E. C. Welsh, memorandum to Douglas Carter, June 12, 1964, LBJPL, White House Central Files, Subject Series—Outer Space, Box 1, File: OS Outer Space 11/22/63–6/19/64, 6.

66. E. C. Welsh, memorandum to the president, May 2, 1964, "Public Reaction to Space," LBJPL, White House Central Files, Subject Series—Outer Space, Box 1, File: OS Outer Space 11/22/63–6/19/64, 1.

67. E. C. Welsh, memorandum to the president, October 24, 1964, "Public Opinion Poll Re Space," LBJPL, White House Central Files, Subject Series-National Aeronautics and Space Council, Box 62, File: FG11–4, 11/22/63–10/31/64, 1.

68. Lyndon Johnson, "Report to the President on the Space Council's Accomplishments," 1961, LBJPL, White House Central Files, Subject Series—Outer Space, Box 1, File: OS Outer Space 11/22/63–6/19/64, 6.

69. Spurgeon Keeny, memorandum to McGeorge Bundy, September 16, 1965, "Public Statement on MOL and Military Uses of Space," LBJPL, National Security File, Intelligence File, Box 10, File: Manned Orbital Laboratory, 1.

70. Reference Research Staff, *President Johnson's Remarks on Disarmament, Arms Control, and Related Subjects 1957–1963,* December 18, 1963, LBJPL, National Security File, Intelligence File, Box 10, File: Manned Orbital Laboratory, 21.

71. Everett C. Dolman, *Astropolitik: Classical Geopolitics in the Space Age,* (London: Frank Cass, 2002), 8.

72. Divine, *The Sputnik Challenge,* xiv–xviii; Walter A. McDougall, *The Heavens and the Earth: A Political History of the Space Age* (Baltimore: Johns Hopkins University Press, 1985), 96, 120; Dickson, *Sputnik: The Shock of the Century,* 224.

73. Reference Research Staff, *President Johnson's Remarks on Disarmament,* 22.

74. Ibid., 20.

75. Ibid. It is interesting to note that by 1962, the Echo communication satellite, which was a passive reflective device, was vastly obsolete compared to the work being done on active commercial communication satellites. For a detailed discussion, see David Whalen, *The Origins of Satellite Communication, 1945–1965* (Washington D.C.: Smithsonian Institution Press, 2002), 19–126.

76. Ed Welsh, memorandum to LBJ, March 26, 1963, "Space Race with the USSR," LBJPL, PLBJ, Vice President Security File, Box 17, File: Space Program-Secret, 1.

77. Ibid.

78. Ibid.

79. Ibid., 1–2.

80. E. C. Welsh, memorandum to LBJ, December 19, 1961, "Space Budget," LBJPL, PLBJ, Vice Presidential Security File, Box 17, File: Space Program-Secret, 1.

81. Robert S. McNamara, memorandum to the president, October 28, 1967, untitled, LBJPL, PLBJ, National Security File, Country File, Box 231, File: USSR-Fractional Orbital Bombardment System, 1. Within this one-page memorandum, McNamara states: "For some time we have observed Soviet Tests consistent with the development of a FOBS. The most recent tests seem to confirm intelligence evidence that the Soviet Union is moving in that direction." This information reached the public through McNamara's news conference of November 3, 1967.

82. Reference Research Staff, *President Johnson's Remarks on Disarmament,* 15.

83. Joseph F. Carrol, memorandum to Joe Califano, May 28, 1964, "Estimate of Soviet Space Spectaculars between Now and the End of 1964," NASAHO, White House, Johnson, Correspondence, File: Johnson Correspondence 1963–1964, 2.

84. Central Intelligence Agency, memorandum, May 30, 1964, "Forecast of Soviet Space Spectaculars in Balance of 1964," NASAHO, White House, Johnson, Correspondence, File: Johnson Correspondence 1963–1964, 1.

85. Chun, *Shooting Down a Star,* 22.

86. Ibid.

87. Ibid.

88. "Secretary's Delegation to the Nineteenth Session of the United Nations General As-

sembly," memorandum of conversation, December 2, 1964, LBJPL, National Security File, Country File, File: USSR 11/64–6/67 Gromyko Conversations, 1.

89. Press Conference Number Forty-Nine of the President of the United States, August 25, 1965, LBJPL, National Security File, Intelligence File, Box 10, File: Manned Orbital Laboratory, 1.

90. McNamara, memorandum to the president, October 28, 1967, 1; Cecil Brownlow, "Soviets Prepare Space Weapons for 1968," *Aviation Week and Space Technology* 108 (November 13, 1967), 30–31; Pavel Podvig, ed., *Russian Strategic Nuclear Forces* (Cambridge: MIT Press, 2001), 196–99.

91. Memorandum of conversation, November 20, 1964, "Disarmament," LBJPL, National Security File, Country File, Europe and USSR, Box 228, File: USSR 11/67–6/67, Gromyko Conversations Vol. 1, 3.

92. Memorandum of conversation, December 9, 1964, U.S.-Soviet Relations, LBJPL, National Security File, Country File, Europe and USSR, Box 228, File: USSR 11/64–6/67, Gromyko Conversations Vol. 1., 2.

93. "Soviet Fractional Orbital Bombardment System," draft press release, undated, LBJPL, National Security File, Country File, Europe and USSR, Box 231, USSR Fractional Orbital Bombardment System, 1.

94. Chester C. Ward, "Space Weapons: U.S. Policy vs. Soviet Achievements," *American Security Council Washington Report,* March 7, 1966, NASAHO, Subjects, Impact Military, 11-F.

95. Cited in ibid., 12-F.

96. Ibid.

97. Howard Simons, "Navy Can Pinpoint Navigation in Any Weather: Secret Satellite System Makes It Possible," *Washington Post,* June 12, 1966, M5; Larry D. Hatfield, "Navigation Satellite Declassified by the Navy," *Washington Post,* July 30, 1967, A31.

98. "The White House Message on the Outer Space Treaty to the United States Senate," February 7, 1967, LBJPL, White House Central Files, Office File of Harry McPherson, Box 16, File: Space, 1.

99. Ibid., 2.

100. Ibid., 3.

101. Transcript of news conference of Secretary of Defense Robert S. McNamara, November 3, 1967, LBJPL, National Security File, Country File, Europe and USSR, Box 231, File: USSR-FOBS, 1.

102. "Statement on FOBS," attached to Spurgeon Keeny memorandum to Walt Rostow, November 9, 1967, LBJPL, National Security File, Country File, Europe and USSR, Box 231, File: USSR-FOBS, 2.

103. Ibid., 3.

104. NASA and USAF, "Summary of Suggestions by NASA Headquarters Personnel as to Ideas That May Have Application to the War in Southeast Asia," December 10, 1964, NASAHO, File: Impact Military, 1–7.

105. Ibid.

Chapter 5

1. These terms are contradictory. However, as John Lewis Gaddis observed in *Strategies of Containment,* each president from Truman to Reagan believed in the idea of containment, but

ultimately chose to pursue the policy with a different emphasis. Likewise, in the evolution of US space policy, Presidents Eisenhower, Kennedy, and Johnson all believed in the same basic objectives, but emphasized different steps to reach these objectives.

2. The RAND Corporation, "Conference on Methods for Studying The Psychological Effects of Unconventional Weapons," Santa Monica, California, January 26–28, 1949, GWUNSA, Microfilm Collection: Militarization of Space, Document 00593, 80–81.

3. Even as late as the 1980s, as the United States and the international community debated the tenets of President's Reagan's Strategic Defense Initiative (SDI), the basic missions of military satellites closely adhered to those identified in 1946.

4. Kenneth Osgood, "Before Sputnik: National Security Policy and the Formation of U.S. Outer Space Policy," in *Reconsidering Sputnik: Forty Years since the Soviet Satellite,* Roger D. Launius, John M. Logsdon, and Robert W. Smith, eds. (New York: Routledge, 2000), 197–229; L. Parker Temple, *Shades of Gray: National Security and the Evolution of Space Reconnaissance* (Reston, Va.: American Institute of Aeronautics and Astronautics, 2005), 91–127.

5. Pavel Podvig, ed., *Russian Strategic Nuclear Forces* (Cambridge: MIT Press, 2001), 196–99; Cecil Brownlow, "Soviets Prepare Space Weapons for 1968," *Aviation Week and Space Technology* 108 (November 13, 1967).

6. Tim Weiner, "Air Force Seeks Bush's Approval for Space Weapons Program," *New York Times,* May 18, 2005, 1; Bryan Bender, "Space Weapons Seen as Possibility," *Boston Globe,* May 19, 2005; Bruce M. Deblois, Richard L. Garwin, R. Scott Kemp, and Jeremy C. Marwell, "Star-Crosses: From Orbiting Lasers to Metal Rods that Strike from the Heavens, the Potential to Wage War from Space Raises Startling Possibilities—and Serious Problems," *IEEE Spectrum,* March 2005, http://spectrum.ieee.org/energy/nuclear/starcrossed (accessed June 3, 2011).

Bibliography

Archives

- Harry S. Truman Presidential Library, Independence, Missouri
  - Papers of Harry S. Truman (Official File; Post Presidential File; SMOF: Psychological Strategy Board Files; Confidential File)
  - Papers of Lawrence R. Hafstad
  - Papers of James E. Webb
  - Papers of K. T. Keller
  - Papers of John H. Ohly
  - Papers of Stuart Symington
  - Papers of Horace E. Weihmiller
  - OF 175: British Fuel Emergency
  - Record Group 220: President's Air Policy Commission
  - Papers of George J. Richards
- National Aeronautics and Space Administration History Office, Washington D.C.
  - Harry S. Truman Papers
  - Dwight D. Eisenhower Papers
  - John F. Kennedy Papers
  - Lyndon Baines Johnson Papers
  - Subject series:
    - SAMOS Satellites
    - USAF Project SAINT
    - Manned Orbiting Laboratory
    - Satellites MIDAS
    - Impact Military
    - Dyna-Soar
    - Blue Gemini
- John F. Kennedy Presidential Library, Boston, Massachusetts
  - Papers of President Kennedy
    - National Security File
      - Department and Agencies
      - Subjects
      - Carl Kaysen Series
    - Pre-Presidential
      - Legislation
  - Papers of John F. Kennedy
    - White House Central Files
      - Subject Series
      - Subject File
- National Security Archive, George Washington University, Washington, D.C.
  - Microfiche Collection
    - The Militarization of Space 1946–1990

Dwight D. Eisenhower Presidential Library, Abilene, Kansas
- Dwight D. Eisenhower: Papers as President
  - Administration Series
  - National Security Council Series
  - DDE Diary Series
  - Legislative Meeting Series
  - Press Conference Series
  - Press Conference Series
  - Speech Series
- Dwight D. Eisenhower: Records as President
  - Official File
- Papers of John Foster Dulles
  - Telephone Conversation Series
- Papers of Clifford Furnas
- Papers of James C. Hagerty
- Sprague Committee Records
- Records of the U.S. President's Advisory Committee
- Papers of the White House Office
  - National Security Council Staff
  - Special Staff Series
- Records of the White House Office
  - Office of the Special Assistant for National Security Affairs
    - NSC Series
      - Briefing Notes Subseries
      - Policy Papers Subseries
      - Subject Subseries
  - Office of the Special Assistant for Science and Technology
    - Records
  - Office of Staff Secretary
    - Subject Series

National Air and Space Museum, Smithsonian Institution, Washington, D.C.
- The RAND Corporation History Project
  - Interview Transcripts
    - Bruno Augenstein
    - Edward Bowles
    - Frank Collbohm
    - Richard Frick
    - Amron Katz
    - Malcolm Palmatier
    - Ernst Plesset
    - Arthur Raymond
    - Bernard Schriever

Lyndon Baines Johnson Presidential Library, University of Texas, Austin, Texas
- White House Central File
  - Aides File
  - Subject File
  - Subject Series
- Papers of Lyndon Baines Johnson
  - Vice Presidential Security File

Senate Papers
President
National Security File
Country File
Subject File
Agency File
Files of Charles E. Johnson
Intelligence File
Files of Spurgeon Kenny

Government Documents and Publications

National Aeronautics and Space Administration. *This New Ocean: A History of Project Mercury.* NASA SP-4201. Washington D.C.: NASA, 1966,.

———. *Project Gemini: A Chronology.* NASA SP-4002Washington D.C.: NASA, 1969.

National Aeronautics and Space Council. *Report to the Congress from the President of the United States: United States Aeronautics and Space Activities, 1965.* Washington D.C.: Government Printing Office, January, 1966.

———. *Report to the Congress from the President of the United States: United States Aeronautics and Space Activities, 1966.* January, 1967.

Report of the Commission to Assess United States National Security Space Management and Organization. Washington D.C.: Government Printing Office, January, 2001.

United Nations. *The United Nations and Disarmament 1945–1970.* New York: United Nations, 1970.

United Nations, Institute for Disarmament Research. *Prevention of an Arms Race in Outer Space: A Guide to the Discussions in the Conference on Disarmament.* New York: United Nations, 1991.

U.S. Arms Control and Disarmament Agency. *Documents on Disarmament 1945–1959: Vol. 1, 1945–1956.* Washington D.C.: Department of State, August 1960.

———. *Documents on Disarmament 1945–1959: Vol. 2, 1957–1959.* Washington D.C.: Department of State, August 1960.

———. *Documents on Disarmament 1960.* Washington D.C.: Department of State, July 1961.

U.S. Congress. Senate. Committee on Aeronautical and Space Sciences. *Manned Space Flight Program of the National Aeronautics and Space Administration: Projects Mercury, Gemini, and Apollo.* Report prepared by Everard H. Smith. 87th Congress., 2d session, September 4, 1962.

United States Information Agency. *President John F. Kennedy Current Statements, 1962.* Washington D.C.: United States Information Agency, 1963.

———. *President John F. Kennedy Current Statements, 1963.* Washington D.C.: United States Information Agency, 1963.

Books

Adams, Chris. *Inside the Cold War: A Cold Warrior's Reflection.* Maxwell Air Force Base: Air University Press, 1999.

The Aerospace Corporation: Its Work 1960–1980. El Segundo, Calif.: Aerospace Corporation, 1980.

Air Force Magazine. *Space Weapons: A Handbook of Military Astronautics.* New York: Frederick A. Praeger, 1959.

Air Force ROTC. *Fundamentals of Aerospace Weapons Systems.* Maxwell Air Force Base: Air University Press, 1961.

Ambrose, Stephen E. *Eisenhower: Soldier and President.* New York: Touchstone, 1990.

Baucom, Donald R. *Origins of SDI, 1944–1983.* Lawrence: University Press of Kansas, 1992.

Berkner, Lloyd V. "The Impact of Sputnik." In *The Impact of Air Power: National Security and World Politics,* ed. Eugene M. Emme, 849–55. Princeton: D. Van Nostrand Company, 1959.

Beschloss, Michael R. *May Day: Eisenhower, Khrushchev and the U-2 Affair.* New York: Harper and Row, 1986.

Biddle, Wayne. *Barons of the Sky from Early Flight to Strategic Warfare: The Story of the American Aerospace Industry.* Baltimore: Johns Hopkins University Press, 1991.

Bille, Matt, and Erika Lishock. *The First Space Race: Launching the World's First Satellite.* College Station: Texas A&M University Press, 2004.

Boushey, Homer A. "Some Military Implications of Astronautics." In *The Impact of Air Power: National Security and World Politics,* ed. Eugene M. Emme, 865–72. Princeton: D. Van Nostrand Company, 1959.

Brodie, Bernard. *Strategy in the Missile Age.* Princeton: Princeton University Press, 1959.

Brookes, Andrew J. *Photo Reconnaissance.* London: Ian Allan, 1975.

Buchheim, Robert W., and the staff of the RAND Corporation. *Space Handbook: Astronautics and Its Applications.* New York: Random House, 1958.

Bulkeley, Rip. *The 'Sputnik' Crisis and the Early United States Space Policy.* Bloomington: Indiana University Press, 1991.

Bulkeley, Rip, and Graham Spinardi. *Space Weapons: Deterrence or Delusion?* Oxford: Polity Press, 1986.

Burrows, William E. *Deep Black: Space Espionage and National Security.* New York: Random House, 1986.

Byrnes, Mark E. *Politics and Space: Image Making by NASA.* Westport: Praeger, 1994.

Chayes, Abram, and Jerome B. Wiesner, eds. *ABM: An Evaluation of the Decision to Deploy an Antiballistic Missile System.* New York: Harper and Row, 1969.

Chun, Clayton K.S. *Shooting Down a "Star": Program 437, the U.S. Nuclear ASAT System and Present-Day Copycat Killers.* CADRE Paper No. 6. Maxwell Air Force Base, Air University Press, 2000.

Cole, Alice C., Alfred Goldberg, Samuel A. Tucker, and Rudolph Winnacker, eds. *The Department of Defense Documents on Establishment and Organization, 1944–1978.* Washington D.C.: Office of the Secretary of Defense, Historical Office, 1978.

Cox, Donald, and Michael Stoiko. *Spacepower: What It Means to You.* Philadelphia: John C. Winston, 1958.

Daso, Dik Alan. *Hap Arnold and the Evolution of American Airpower.* Washington D.C.: Smithsonian Institution Press, 2000.

———. *Architects of American Air Supremacy: General Hap Arnold and Dr. Theodore von Kármán.* Maxwell Air Force Base: Air University Press, 1977.

Day, Dwayne A., John M. Logsdon, and Brian Latell, eds. *Eye in the Sky: The Story of the Corona Spy Satellite.* Washington D.C.: Smithsonian Institution Press, 1993.

DeVorkin, David H. *Science with a Vengeance: How the Military Created the US Space Sciences after World War II.* Reprint. New York: Springer-Verlag, 1993.

Diamond, Edwin. *The Rise and Fall of the Space Age.* Garden City: Doubleday, 1964.

Dickson, Paul. *Sputnik: The Shock of the Century.* New York: Walker, 2001.

Divine, Robert A. *The Sputnik Challenge.* New York: Oxford University Press, 1993.

Dolman, Everett C. *Astropolitik: Classical Geopolitics in the Space Age.* London: Frank Cass, 2002.

Durant, Frederick C., III, ed. *Between Sputnik and the Shuttle: New Perspectives on American Astronautics.* San Diego: American Astronautical Society, 1981.

Emme, Eugene M., ed. *The Impact of Air Power: National Security and World Politics.* New York: D. Van Nostrand Company, 1959.

Farquhar, John Thomas. *A Need to Know: The Role of Air Force Reconnaissance in War Planning, 1945–1953.* Maxwell Air Force Base: Air University Press, 2004.

Friedman, Norman. *Seapower and Space: From the Dawn of Missile Age to Net-Centric Warfare.* Annapolis: Naval Institute Press, 2000.

Gaddis, John Lewis. *Strategies of Containment: A Critical Appraisal of Postwar American National Security Policy.* Oxford: Oxford University Press, 1982.

Gantz, Kenneth F., ed. *Man in Space: The United States Air Force Program for Developing Spacecraft Crew.* New York: Duell, Sloan and Pearce, 1959.

———. *The United States Air Force Report on the Ballistic Missile: Its Technology, Logistics, and Strategy.* New York: Doubleday, 1958.

Godwin, Robert, ed. *Dyna-Soar: Hypersonic Strategic Weapons System.* Ontario: Apogee Books, 2003.

Golovine, M. N. *Conflict in Space: A Pattern of War in a New Dimension.* New York: St. Martin's Press, 1962.

Gonzales, Daniel. *The Changing Role of the U.S. Military in Space.* Santa Monica: RAND Corporation, 1999.

Gorn, Michael H. *Harnessing the Genie: Science and Technology Forecasting for the Air Force, 1944–1986.* Washington D.C.: United States Air Force, Office of Air Force History, 1988.

Graham, Thomas, and Damien J. LaVera. *Cornerstones of Security: Arms Control Treaties in the Nuclear Era.* Seattle: University of Washington Press, 2003.

Gray, Colin S. *The Leverage of Sea Power: The Strategic Advantage of Navies in War.* New York: Free Press, 1992.

———. *The Geopolitics of Super Power.* Lexington: University Press of Kentucky, 1988.

———. *American Military Space Policy: Information Systems, Weapons Systems, and Arms Control.* Cambridge: ABT Books, 1982.

Grossman, Karl. *Weapons in Space.* New York: Seven Stories Press, 2001.

Hall, R. Cargill, and Jacob Neufeld., eds. *The U.S. Air Force in Space: 1945 to the Twenty-first Century.* Washington D.C.: USAF History and Museums Program, 1988.

Hallion, Richard P. *The Literature of Aeronautics, Astronautics, and Air Power.* Washington D.C.: United States Air Force, Office of Air Force History, 1984.

Hartford, James. *Korolev: How One Man Masterminded the Soviet Drive to Beat America to the Moon.* New York: John Wiley and Sons, 1997.

Hattendorf, John B., and Robert S. Jordan, eds. *Maritime Strategy and the Balance of Power: Britain and America in the Twentieth Century.* New York: St. Marin's Press, 1989.

Hensell, Howard M. *The U.S.S.R. and Outer Space: Media Images of Superpower Space Policy.* Montgomery, Ala.: Air Command and Staff College, 1985.

Heppenheimer, T. A. *Countdown: A History of Space Flight.* New York: John Wiley and Sons, 1997.

Holley, I. B., Jr. *Ideas and Weapons Exploitation of the Aerial Weapon by the United States during*

World War I: A Study in the Relationship of Technological Advance, Military Doctrine, and the Development of Weapons. Hamden, Conn.: Archon Books, 1971.

Humble, Ronald D. *The Soviet Space Programme.* London: Routledge, 1988.

Hunley, J. D., ed. *The Birth of NASA: The Diary of T. Keith Glennan.* Washington D.C.: NASA History Office, 1993.

Jasani, Bhupendra, ed. *Outer Space: A New Dimension of the Arms Race.* London: Taylor and Francis, 1982.

Jockel, Joseph T. *No Boundaries Upstairs: Canada, the United States, and the Origins of North American Air Defence, 1945–1958.* Vancouver: University of British Columbia, 1987.

Johnson, Dana J., Scott Pace, and C. Bryan Gabbard. *Space: Emerging Options for National Power.* Santa Monica: RAND Corporation, 1998.

Johnson, Stephen B. *The United States Air Force and the Culture of Innovation, 1945–1965.* Washington D.C.: Air Force History and Museums Program, 2002.

Kirby, Stephen, and Gordon Robson., eds. *The Militarization of Space.* Sussex: Wheatsheaf Books, 1987.

Kistiakowsky, George B., and Charles S. Maier. *A Scientist at the White House: The Private Diary of President Eisenhower's Special Assistant for Science and Technology.* Cambridge: Harvard University Press, 1976.

Komons, Nick A. *Science and the Air Force: A History of the Air Force Office of Scientific Research.* Arlington, Va.: Historical Division, Office of Information, Office of Aerospace Research, 1966.

Krug, Linda T. *Presidential Perspectives on Space Exploration: Guiding Metaphors from Eisenhower to Bush.* New York: Praeger, 1991.

Lambeth, Benjamin S. *Mastering the Ultimate High Ground: Next Steps in the Military Uses of Space.* Santa Monica: RAND Corporation, 2003.

Lang, Daniel. *From Hiroshima to the Moon: Chronicles of Life in the Atomic Age.* New York: Simon and Schuster, 1959.

Lashmar, Paul. *Spy Flights of the Cold War.* Gloucestershire: Sutton Publishing, 1996.

Launius, Roger D., John M. Logsdon, and Robert W. Smith, eds. *Reconsidering 'Sputnik': Forty Years since the Soviet Satellite.* London: Routledge, 2000.

Launius, Roger D., and Howard E. McCurdy, eds. *Spaceflight and the Myth of Presidential Leadership.* Urbana: University of Illinois Press, 1977.

Lee, Christopher. *War in Space.* London: Hamish Hamilton, 1986.

Levine, Alan J. *The Missile and the Space Race.* Westport: Praeger, 1994.

Ley, Willy. *Rockets, Missiles, and Space Travel.* New York: Viking Press, 1958.

Lockwood, Jonathan Samuel, and Kathleen O'Brien Lockwood. *The Russian View of U.S. Strategy: Its Past, Its Future.* New Brunswick: Transaction Publishers, 1993.

Logsdon, John M, Dwayne A. Day, and Roger D. Launius, eds. *Exploring the Unknown: Selected Documents in the History of the U.S. Civil Space Program, Volume II, External Relationships.* Washington D.C.: NASA, History Office, 1996.

Long, Franklin A., Donald Hafner, and Jeffrey Boutwell, eds. *Weapons in Space.* New York: W. W. Norton, 1986.

Loosbrock, John F., and Richard M. Skinner, eds. *The Wild Blue: The Story of American Airpower.* New York: G. P. Putnam's Sons, 1961.

Lupton, David E. *On Space Warfare: A Space Power Doctrine.* Maxwell Air Force Base: Air University Press, 1988.

MacCloskey, Monroe. *From Gasbags to Spaceships: The Story of the U.S. Air Force.* New York: Richard Rosen Press, 1968.

Manno, Jack. *Arming the Heavens: The Hidden Military Agenda for Space, 1945–1995.* New York: Dodd and Mead, 1984.

McCurdy, Howard E. *Space and the American Imagination.* Washington D.C.: Smithsonian Institution Press, 1985.

McDougall, Walter A. *The Heavens and the Earth: A Political History of the Space Age.* Baltimore: Johns Hopkins University Press, 1985.

Mead, Peter. *The Eye in the Air: History of Air Observation and Reconnaissance for the Army 1785–1945.* London: Her Majesty's Stationery Office, 1983.

Meilinger, Philip S., ed. *The Paths of Heaven: The Evolution of Air Power Theory.* Maxwell Air Force Base: Air University Press, School of Advanced Air Power Studies, 1997.

Mowthorpe, Matthew. *The Militarization and Weaponization of Space.* New York: Lexington Books, 2004.

Muolo, Michael J. *Space Handbook: A War Fighter's Guide to Space.* Maxwell Air Force Base: Air University Press, 1993.

Myers, Grover E. *Aerospace Power: The Case for Indivisible Application.* Maxwell Air Force Base: Air University Press, 1986.

Neufeld, Jacob. *Development of Ballistic Missiles in the United States Air Force, 1945–1960.* Washington D.C.: United States Air Force, Office of Air Force History, 1990.

———, ed. *Reflections on Research and Development in the United States Air Force: An Interview with General Bernard A. Schriever, and Generals Samuel C. Phillips, Robert T. Marsh, and James H. Doolittle, and Dr. Ivan A. Getting.* Washington D.C.: Center for Air Force History, 1993.

Neufeld, Jacob, George M. Watson, Jr., and David Chenoweth. *Technology and the Air Force: A Retrospective Assessment.* Washington D.C.: United States Air Force, Air Force History and Museums Program, 1997.

Newell, Homer E. *Sounding Rockets.* New York: McGraw-Hill, 1959.

Oberg, James E. *Red Star in Orbit.* New York: Random House, 1981.

Oberth, Hermann. *Man into Space: New Projects for Rocket and Space Travel.* New York: Harper and Brothers, 1957.

O'Leary, Michael. *U.S. Sky Spies since World War I.* Poole: Blandford Press, 1986.

Ordway, Frederick I., and Ronald C. Wakeford. *International Missile and Spacecraft Guide.* New York: McGraw-Hill, 1960.

Ossenbeck, Frederick J., and Patricia C. Kroeck, eds. *Open Space and Peace.* Stanford: Hoover Institution, 1964.

Patch, Chester J., and Elmo Richardson. *The Presidency of Dwight D. Eisenhower.* Lawrence: University Press of Kansas, 1991.

Peebles, Curtis. *The Corona Project: America's First Spy Satellite.* Annapolis: Naval Institute Press, 1997.

———. *High Frontier: The U.S. Air Force and the Military Space Program.* Washington D.C.: Air Force History and Museums Program, 1997.

Pedlow, Gregory W., and Donald E. Welzenbach. *The CIA and the U-2 Program.* Central Intelligence Agency, History Staff, Center for the Study of Intelligence, 1998.

Perminov, V.G. *The Difficult Road to Mars: A History of Mars Exploration in the Soviet Union.* Translated by Katherine A. Navarova. Washington D.C.: NASA, 1999.

Piszkiewicz, Dennis. *The Nazi Rocketeers: Dreams of Space and Crimes of War.* Westport: Praeger, 1995.

Preston, Bob, Dana J. Johnson, Sean J. A. Edwards, Michael Miller, and Calvin Shipbaugh. *Space Weapons: Earth Wars.* Santa Monica: RAND Corporation, 2002.

The RAND Corporation: The First Fifteen Years. Santa Monica: RAND Corporation, 1963.

Richelson, Jeffrey T. *America's Space Sentinels: DSP Satellites and National Security.* Lawrence: University Press of Kansas, 1999.

Ritchie, David. *Space War.* New York: Atheneum, 1982.

Rosen, Milton W. *The Viking Rocket Story.* New York: Harper, 1955.

Rostow, Eugene V. *Toward Managed Peace: The National Security Interests of the United States, 1759 to the Present.* New Haven: Yale University Press, 1993.

Rostow, W. W. *Open Skies: Eisenhower's Proposal of July 21, 1955.* Austin: University of Texas Press, 1982.

Ruffner, Kevin C., ed. *Corona: America's First Satellite Program.* Washington D.C.: Center for the Study of Intelligence, 1995.

Salkeld, Robert. *War and Space.* Englewood Cliffs, N.J.: Prentice-Hall, 1970.

Sapolosky, Harvey M. *The Polaris System Development: Bureaucratic and Programmatic Success in Government.* Cambridge: Harvard University Press, 1972.

Sharpe, Mitchell R. *Satellites and Probes: The Development of Unmanned Space Flight.* Garden City, N.J.: Doubleday, 1970.

Sheldon, Charles S. *Review of the Soviet Space Program with Comparative United States Data.* New York: McGraw-Hill, 1986.

Siddiqi, Asif A. *Challenge to Apollo: The Soviet Union and the Space Race, 1945–1974.* Washington D.C.: National Aeronautics and Space Administrations, History Division, Office of Policy and Plans, 2000.

Smolders, Peter L. *The Soviets in Space: The Story of the Salyut and the Soviet Approach to Present and Future Space Travel.* Translated by Marian Powell. London: Lutterworth Press, 1973.

The Space Industry: America's Newest Giant. Englewood Cliffs: Prentice Hall, 1962.

Spires, David N. *Beyond Horizons: A Half Century of Air Force Leadership in Space.* Maxwell Air Force Base: Air University Press, 1998.

Spykman, Nicholas John. *America's Strategy in World Politics: The United States and the Balance of Power.* New York: Harcourt, Brace, 1942.

Stares, Paul B. *The Militarization of Space: U.S. Policy 1945–1984.* Ithaca: Cornell University Press, 1985.

———. *Space Weapons and U.S. Strategy: Origins and Development.* London: Croom Helm, 1985.

Steinberg, Gerald M. *Satellite Reconnaissance: The Role of Informal Bargaining.* New York: Praeger, 1983.

Stine, G. Harry. *ICBM: The Making of the Weapon that Changed the World.* New York: Orion Books, 1991.

Straubel, James H. et al., eds. *Space Weapons: A Handbook of Military Astronautics.* New York: Praeger, 1959.

Stumpf, David K. *Titan II: A History of a Cold War Missile Program.* Fayetteville: University of Arkansas Press, 2000.

Swenson, Loyd S. Jr., James M. Grimwood, and Charles C. Alexander. *This New Ocean: A History of Project Mercury.* Washington D.C.: NASA, Office of Technology Utilization, 1966.

Taubman, Philip. *Secret Empire: Eisenhower, the CIA, and the Hidden Story of America's Space Espionage.* New York: Simon and Schuster, 2003.

Terrill, Delbert R. *The Air Force Role in Developing International Outer Space Law.* Maxwell Air Force Base: Air University Press, May 1999.

Van Dyke, Vernon. *Pride and Power: The Rationale of the Space Program.* Urbana: University of Illinois Press, 1964.

Von Karman, Theodore, with Lee Edson. *The Wind and Beyond: Theodore Von Karman, Pioneer in Aviation and Pathfinder in Space.* Boston: Little, Brown, 1967.

Walker, James A., Frances Martin, and Sharon S. Watkins. *Strategic Defense: Four Decades of Progress.* U.S. Army Space and Strategic Defense Command, History Office, 1995.

Watson, George M. *The Office of the Secretary of the Air Force, 1947–1965.* Washington D.C.: Center for Air Force History, 1993.

Whalen, David J. *The Origins of Satellite Communications, 1945–1965.* Washington D.C.: Smithsonian Institution Press, 2002.

White, T. D., and B. A. Shriever. "Immediate Prospects in Astronautics." In *The Impact of Air Power: National Security and World Politics,* ed. Eugene M. Emme, 856–64. Princeton: D. Van Nostrand Company, 1959.

Winter, Frank H. *Prelude to the Space Age: The Rocket Societies 1924–1940.* Washington D.C.: Smithsonian Institution Press, 1983.

Reports, Conference Papers, Dissertations, and Theses

Bell, Thomas D. *Weaponization of Space: Understanding Strategic and Technological Inevitabilities.* Maxwell Air Force Base: Air War College, Center for Strategy and Technology, January 1999, Occasional Paper No. 6.

Borowski, Lisa M. "Military Priorities in the Cold War: Operation ARGUS, 1957–1959." Master's thesis, Kansas State University, May 2000.

Brodie, Bernard. *The Heritage of Douhet.* Santa Monica: RAND Corporation, December 21, 1952, RM-1013.

Chisholm, Robert H. *On Space Warfare: Military Strategy for Space Operations.* Maxwell Air Force Base: Air University Press, June 1984, AU-ARI-84–3.

Davies, Merton E., and William R. Harris. *RAND's Role in the Evolution of Balloon and Satellite Observation Systems and Related U.S. Space Technology.* Santa Monica: RAND Corporation, September 1988, R-3692-RC.

Frye, William E. ed. *Impact of Space Exploration on Society: Proceedings of the National Meeting of the American Aeronautical Society 18–20 August 1965.* Tarzana, Calif.: American Astronautical Society, 1966.

Goldsen, Joseph M. *Outer Space and the International Scene.* Santa Monica: RAND Corporation, May 6, 1969, P-1688.

Goldsen, Joseph M., and Leon Lipson. *Some Political Implications of the Space Age.* Santa Monica: RAND Corporation, February 24, 1958, P-1435.

Hagan, Regina. "(Un-)Peaceful Uses of Space." Presentation for the 13th General Assembly of the International Association of Peace Messenger Cities, Oswiecim, Poland, September 3, 2000.

Hall, R. Cargill and Jacob Neufeld, eds. *The U.S. Air Force in Space: 1945 to the Twenty-first Century.* Proceedings, Air Force Historical Foundation Symposium, Andrews AFB, Maryland, September 21–22, 1995. Washington D.C.: United States Air Force, USAF History and Museums Program, 1998.

Hays, Peter Lang. "Struggling towards Space Doctrine: U.S. Military Space Plans, Programs, and Perspectives During the Cold War." PhD diss., Fletcher School of Law and Diplomacy, Tufts University, 1994.

———. *United States Military Space: Into the Twenty-First Century.* Colorado Springs: USAFA Institute for National Security Studies, September 2002, INSS Occasional Paper 42.

Heitchue, R. D. *Space Age Fundamentals.* Douglas Missile and Space Systems Division: Douglas Air Craft Company, July 1964, SM-47656.

Heymann, Hans. *The USSR in the Technological Race.* Santa Monica: RAND Corporation, July 20, 1959, P-1754.

Hitchcock, Water T., ed. *The Intelligence Revolution: A Historical Perspective: Proceedings of the Thirteenth Military History Symposium U.S. Air Force Academy, Colorado Springs, October 12–14, 1988.* Washington D.C.: Office of Air Force History, 1991.

Hoff, Nicholas John, and Walter Guido Vincenti., eds. *Aeronautics and Astronautics: Proceedings of the Durand Centennial Conference Held at Stanford University, 5–8 August 1959.* New York: Pergamon Press, 1960.

Johnson, Dana Joyce. "The Evolution of U.S. Military Space Doctrine: Precedents, Prospects, and Challenges." PhD diss., University of Southern California, December 1987.

Kalic, Sean N. "Withering Triad: An Examination of the Modernization Programs for the Strategic Nuclear Forces of the United States." Master's thesis, Southwest Missouri State University, May 1997.

Krieger, F. J. *A Casebook on Soviet Astronautics.* Santa Monica: RAND Corporation, June 21, 1956, RM-1760.

———. *Casebook on Soviet Astronautics—Part II.* Santa Monica: RAND Corporation, June 21, 1957, RM-1922.

———. *Recent Soviet Advances in Aerospace Technology.* Santa Monica: RAND Corporation, February 1962, RM-3053-PR.

MacDonald, Steven D. "Existing and Prospective International Legal Restrictions on the Military Uses of Outer Space." LL.M. thesis, University of Ottawa, February 1990.

Mantz, Michael R. *The New Sword: A Theory of Space Combat Power.* Maxwell Air Force Base: Air University Press, May 1995, AU-ARI-94–6.

President's Science Advisory Committee. *The Space Program in the Post-Apollo Period.* Washington D.C.: Government Printing Office, February 1967.

RAND Corporation. *The Preliminary Design of an Experimental World-Circling Spaceship.* Reprint. Santa Monica: RAND Corporation, 1946, SM-11827.

Rothstein, Stephen M. "Dead on Arrival? The Development of the Aerospace Concept, 1944–1958." MMAS thesis, School of Advanced Airpower Studies, November 2000.

Tucker, Robert C. *The Psychological Factor in Soviet Foreign Policy.* Santa Monica: RAND Corporation, March 7, 1957, RM-1881.

Articles and Electronic Sources

Bender, Bryan. "Space Weapons Seen as Possibility: New Policy Expected to Exceed Guidelines Emphasizing Defense." *Boston Globe* May 19, 2005.

Boushey, Homer A. "The USAF and Astronautical Development." *Airpower Historian* 5 (April 1958): 118–23.

Brownlow, Cecil. "Soviets Prepare Space Weapons for 1968." *Aviation Week and Space Technology* 108 (November 13, 1967): 30–31.

Cooper, John Cobb. "Aerospace Law—Recent Development." *Airpower Historian* 10 (October 1963): 118–20.

Deblois, Bruce. "Space Sanctuary: A Viable National Strategy." *Airpower Journal* 12 (Winter 1998): 41–57.

Deblois, Bruce, Richard L. Garwin, R. Scott Kemp, and Jeremy C. Marwell. "Star-Crosses:

From Orbiting Lasers to Metal Rods that Strike from the Heavens, the Potential to Wage War from Space Raises Startling Possibilities—and Serious Problems." *IEEE Spectrum,* March 2005, http://spectrum.ieee.org/energy/nuclear/starcrossed (accessed June 3, 2011).

———. "Space Weapons: Crossing the U.S. Rubicon." *International Security* 14 (Fall 2004): 50–84.

De Seversky, Alexander. "Air Power, Missiles, and National Survival." *Airpower Historian* 5 (January 1958): 20–30.

———. "Command of Space." *Airpower Historian* 9 (October 1962): 253.

Dryden, Hugh L. "The National Significance of the Augmented Program of Space Exploration." *Airpower Historian* 9 (January 1962): 52–59.

Gardner, Trevor. "How We Fell Behind in Guided Missiles." *Airpower Historian* 5 (January 1958): 3–13.

Getting, Ivan A. "The Space Dichotomy." *Airpower Historian* 9 (October 1962): 225–30.

Goldwater, Barry M. "America in Space." *Airpower Historian* 10 (April 1963): 43–45.

Heintze, Hans-Joachim. "Peaceful Uses of Outer Space and International Law," Global Network Against Weapons and Nuclear Power in Space, http://space4peace.org/ethics/puosil.htm (July 29, 2005).

Institute of Air and Space Law. "'Peaceful' and Military Uses of Outer Space" Law and Policy." McGill University, February 2005.

LeMay. Curtis E. "Our National Position in the World from a Military Point of View." *Airpower Historian* 9 (July 1962): 134–40.

MacFarland, Marvin. "Men of Space." *Airpower Historian* 10 (October 1963): 127.

Mueller, Karl P. "Totem and Taboo: Depolarizing the Space Weaponization Debate." *Astropolitics* 1 (Spring 2003): 4–28.

Neufeld, Michael J. "The End of the Army Space Program: Inter-service Rivalry and the Transfer of the von Braun Group to NASA," *Journal of Military History* 69 (July 2005): 735–57.

Pépin, E. "Space Penetration." *Airpower Historian* 4 (October 1960): 233–38.

Stanton, Thomas F. "The Battle for Men's Minds." *Airpower Historian* 7 (April 1960): 81–92.

———. "The Psycho-Social Instrument of National Policy." *Airpower Historian* 8 (October 1961): 207–19.

United States Arms Control and Disarmament Agency. "Early U.S. Arms Control and Disarmament Initiatives." http://usinfo.state.gov/products/pubs/archive/armscrtl/pts3.htm (accessed July 29, 2005).

Vaughn, Chris. "General: Weapons Not Goal." *Star-Telegram* July 28, 2005.

White, Thomas D. "At the Dawn of the Space Age." *Airpower Historian* 5 (January 1958): 15–20.

———. "Leadership in the Conquest of Space." *Airpower Historian* 5 (April 1958): 75–80.

Index

Page numbers in *italics* indicate images.

Other titles in the Centennial of Flight Series

America's Airports Janet R. Daly Bednarek

American Military Aviation Charles J. Gross

Like Sex with Gods: An Unorthodox History of Flying Bayla Singer

Dreams of Flight Bednarek and Bednarek

100 Years of Air Power and Aviation Robin Higham

The Bird Is on the Wing: Aerodynamics and the Progress of the American Airplane James R. Hansen

Imagining Flight: Aviation and Popular Culture A. Bowdoin Van Riper

The First Space Race: Launching the World's First Satellites Bille and Lishock

Electronics in the Evolution of Flight Albert Helfrick

Flying Down to Rio: Hollywood, Tourists, and Yankee Clippers Rosalie Schwartz

Sky as Frontier: Adventure, Aviation, and Empire David T. Courtwright

Spying from Space Arnold and McCartney

Into the Wild Blue Yonder: Allan T. Stein

From the Pilot Factory, 1942 William P. Mitchell

Testing the Limits: Aviation Medicine and the Origins of Manned Space Flight Maura Phillips Mackowski

Aviator of Fortune: Lowell Yerex and the Anglo-American Commercial Rivalry, 1931–1946 Erik Benson

Development of Propulsion Technology for U.S. Space-Launch Vehicles, 1926–1991 J. D. Hunley

Red Cosmos: K. E. Tsiolkovskii, Grandfather of Soviet Rocketry James T. Andrews